MOBILITY AND PL

Mobility and Place
Enacting Northern European Peripheries

Edited by
JØRGEN OLE BÆRENHOLDT
Roskilde University, Denmark

BRYNHILD GRANÅS
University of Tromsø, Norway

Department of Planning and Community Studies, University of Tromsø

Routledge
Taylor & Francis Group

LONDON AND NEW YORK

First published 2008 by Ashgate Publishing

Published 2016 by Routledge
2 Park Square, Milton Park, Abingdon, Oxfordshire OX14 4RN
711 Third Avenue, New York, NY 10017, USA

First issued in paperback 2016

Routledge is an imprint of the Taylor & Francis Group, an informa business

British Library Cataloguing in Publication Data
Mobility and place : enacting Northern European peripheries
  1. Social change - Europe, Northern 2. Ethnicity - Europe,
  Northern 3. Europe, Northern - Social conditions - 21st
  century
  I. Bærenholdt, Jørgen Ole II. Granås, Brynhild
  303.4'0948

Library of Congress Cataloging-in-Publication Data
Mobility and place : enacting Northern European peripheries / edited by Jørgen Ole
Bærenholdt and Brynhild Granås.
     p. cm.
  Includes index.
  ISBN 978-0-7546-7141-1
  1. Migration, Internal--Europe, Northern. 2. Cultural landscapes--Europe, Northern. 3.
Cultural geography--Europe, Northern. I. Bærenholdt, Jørgen Ole, 1959- II. Granås,
Brynhild, 1970-

  HB2041.A3M63 2008
  304.80948--dc22

                                                                              2007046583

Cover image: *Horizon* by Arvid Sveen (photo and copyright). Picture from Nykvåg in Vesterålen, Norway.

ISBN 13: 978-1-138-27011-4 (pbk)
ISBN 13: 978-0-7546-7141-1 (hbk)

Contents

Part 3 Mobilizing Place

List of Figures

List of Contributors

Jørgen Ole Bærenholdt MA in history and geography, PhD geography, Dr. Scient. Soc. in social sciences. Associate Professor in Geography at Roskilde University, Department of Environmental, Social and Spatial Change (ENSPAC), Roskilde University. Has been a Professor in the Department of Planning and Community Studies at University of Tromsø. Research in culture, regional development and tourism, in the North Atlantic and Denmark. Co-ordinator of UNESCO MOST CCPP 1996–2001. Among his books in English are *The Reflexive North* (2001, edited with Nils Aarsæther), *Performing Tourist Places* (2004, with Michael Haldrup, Jonas Larsen and John Urry), *Space Odysseys* (2004, edited with Kirsten Simonsen) and *Coping with Distances, Producing Nordic Atlantic Societies* (2007, habil. thesis).

Karl Benediktsson BS in geography from University of Iceland, MA in geography from University of Auckland, PhD in human geography from the Australian National University, Canberra. Professor in Human Geography, University of Iceland. Research in development and rural spaces in Papua New Guinea as well as post-productivist agriculture, coping processes, landscape, nature and place re-invention in Iceland. Among his publications are international journal articles in *Sociologia Ruralis* and *Geografiska Annaler* and the monograph *Harvesting Development: The Construction of Fresh Food Markets in Papua New Guinea* (2002).

Inger Birkeland MA and PhD in geography from University of Oslo. Has had positions at several university colleges in Norway and now holds a Post-doc position in the Department of Geography at the University of Bergen. Research in cultural and feminist geography, tourism and travel, environmental justice, and place planning. Among her publications is *Making Place, Making Self: Travel, Subjectivity and Sexual Difference* (2005).

Marianne Brekke MA in planning from University of Tromsø with thesis on globalization and young people in India. PhD student in the Department of Planning and Community Studies as part of the 'Globalization from Below' project; a Social Science Faculty priority project at the University of Tromsø with support from the Research Council of Norway. PhD project on young refugees' lives in Tromsø and transnational relations.

Anne Britt Flemmen MA in geography from University of Trondheim, PhD in sociology, University of Tromsø. Associate Professor in the Department of Sociology at University of Tromsø. Research on gender issues both in Ethiopia and Northern Norway, including gender policies and marriages between Russian migrant women and Norwegian men. She has published very many book contributions and articles, for example in *Social Semiotics*. Member of steering committee for the Research School 'Citizenship, Encounters and Place Enactment' (CEPIN) at the University of Tromsø.

Siri Gerrard   Magistergrad in ethnography/social anthropology from University of Tromsø. Associate Professor in the Department of Planning and Community Studies at the University of Tromsø. Many years of research experience on gender, work, resources and community life in Northern Norway and Africa. Research co-operation with among others Memorial University, St. Johns. In addition to national and international journal articles, she has co-edited *Global Coasts: Life Changes, Gender Challenges* (1999, edited with Randi Rønning Balsvik), *Changing Tides, Gender, Fisheries and Globalization* (2005, edited with Barbara Neis, Marian Binkley and Maria Cristina Maneschy) and *Challenging Situatedness, Gender, Culture and the Production of Knowledge* (2005, edited with Ericka Engelstad).

Brynhild Granås   MA in sociology from University of Oslo. PhD student in the Department of Sociology, participant at the Research School 'Citizenship, Encounters and Place Enactment' (CEPIN), University of Tromsø. Until 2007 Research Officer in the Department of Planning and Community Studies, managing the 'Globalization from Below' project; a Social Science Faculty priority project at the University of Tromsø with support from the Research Council of Norway, where she developed the unique competence of combining research management with being an active and publishing researcher in several international research projects. Among her publications is *Place Reinvention in the North. Dynamics and Governance Perspectives* (2007, edited with Torill Nyseth).

Willy Guneriussen   Dr.philos. in philosophy, Professor in the Department of Sociology at University of Tromsø, also acting as the research director of Social Science Faculty and member of the steering committee of the Research School 'Citizenship, Encounters and Place Enactment' (CEPIN). Visiting Professor at Harstad University College. Research in fundamentals of social science and in philosophy of science. Among his many publications are the books *Aktør, Handling og Struktur* (1996, in Swedish 1997, second edition 1999) and *Å forstå det Moderne* (1999).

Gestur Hovgaard   MA in administration from Roskilde University. PhD in social sciences from Roskilde University. PhD thesis *Globalisation, Embeddedness and Local Coping Strategies*. Senior researcher and acting managing director of the Research Centre on Regional and Local Development in the Faroes. Now Assistant Professor at Roskilde University. Research in economic sociology, innovation, youth identity and local development, and also an entrepreneur in Faroese bio-technology. Published *The Restructuration of the Faroese Economy, The Significance of the Inner Periphery* (2001, with Richard Apostle et al.).

Tim Ingold   First degrees and PhD from Cambridge University in social anthropology. Professor in Social Anthropology, Aberdeen University, earlier at University of Manchester. Has published extensively on Northern Circumpolar people, their movement in the landscape, and especially about Sámi and Finns in Northern Finland. Among his books are *The Skolt Lapps Today* (1976), *Hunters, Pastoralists and Ranchers* (1980), *The Appropriation of Nature* (1986), *Evolution of Social Life* (1986), *Key Debates in Anthropology* (1996, ed.), *The Perception of the Environment, Essays on Livelihood, Dwelling and Skill* (2000) and *Lines: A Brief History* (2007).

Johan Jansson   MA and PhD in economic geography. Research Fellow at the Department of Social and Economic Geography, Uppsala University. PhD Thesis *The Internet Industry in Central Stockholm – A Study of Agglomeration Economies, Social Network Relations and Information Flows* (in Swedish). Main research field in the economic geography of cultural industries, agglomerations, creative milieus, entrepreneurship and firm location.

Gunnar Thór Jóhannesson   BA and MA in anthropology from the University of Iceland. PhD in geography from Roskilde University. Project Manager at the University of Iceland Social Science Research Institute. Research in rural transformation, tourism, innovation and entrepreneurship, destination development and place in Iceland and in the Faroes, with a point of departure in Actor Network Theory. PhD thesis: *Emergent Tourism: An Actor-Network Approach to Tourism Economies* (2007). International articles published in *Sociologia Ruralis* and *Tourist Studies*.

Siv Ellen Kraft   MA in religious studies from the University of Tromsø. PhD in the history of religion, University of Bergen. PhD thesis *The Sex Problem. Political Aspects of Gender Discourse in the Theosophical Society 1875–1930*. Associate Professor in Religious Studies at University of Tromsø. Has also been a lecturer at the University of Bergen. Research in sex, gender, and modification of the body. She has published very many articles and contributions.

Samal Matras Kristiansen   MA in geography and social policy, with a specialization in urban studies, from Roskilde University 2005 with thesis on mobility and housing in the Faroes. Director of Communication, the University of the Faroes. Teaching in High School and also working as a consultant. Written papers to PhD courses, participated in the 'Globalization from Below' project, written many newspaper comments.

Jonas Larsen   BA in social policy, Roskilde Univeristy, MA in cultural studies, Lancaster University, PhD in cultural geography, Roskilde University. PhD-thesis *Performing Tourist Photography*. Associate Professor in Sociology, Aalborg University. Has been an Assistant Professor in Geography at Roskilde University, and research fellow at the Centre of Mobility Research (CeMoRe) at Lancaster University. In addition to several journal articles, his books include *Performing Tourist Places* (2004, with Jørgen Ole Bærenholdt, Michael Haldrup and John Urry) and *Mobilities, Networks, Geographies* (2006, with John Urry and Kay Axhausen).

Ari Aukusti Lehtinen   PhD in geography from the University of Helsinki and Professor in Geography, University of Joensuu. Earlier positions at the University of Helsinki and co-operation with other Nordic universities. Research in biopolitics, environmental justice, especially in relation to forestry. Among his many publications are *Northern Natures. A Study of the forest question emerging within the timber-line conflict* (1991), *Politics of Forests* (2004, edited with Jacob Donner-Amnell and Bjørnar Sæther) and *Postcolonialism, Multitude, and the Politics of Nature, On the Changing Geographies of the European North* (2006).

Ann Therese Lotherington   Political scientist, MA from University of Oslo and PhD from University of Tromsø. Research Chair in NORUT, Tromsø. Research in gender, power, migration and regional development. Numerous publications and participation in international projects in the Arctic. Member of steering committee for the Research School 'Citizenship, Encounters and Place Enactment' (CEPIN) at the University of Tromsø.

Nuccio Mazzullo   Degree in sociology at the University of Urbino, MA and PhD social anthropology, University of Manchester. PhD-thesis *Perception, Tradition and Environment among Sámi people in Northeastern Finland*. Research associate in the anthropology research team at the Arctic Centre, University of Lapland, Rovaniemi. Topical interests include environmental politics and indigenous rights, perception of landscape, orientation and place naming, learning and skills and anthropology of circumpolar peoples. His recent publication is 'Environmental Conservation and Local Interests in Finnish Lapland', in *Conservation and Society* (vol. 3, Number 2, 2005).

Torill Nyseth   MA and PhD in planning. Associate Professor in the Department of Planning and Community Studies, University of Tromsø. Research in local democracy, public policy, governance and administration and in urban development and design. She has also been a college lecturer and worked as a municipal planner. In addition to many journal articles and contributions in political science, entrepreneurship and rural development, she has published the book *Dugnad og Demokrati* (2000, with Toril Ringholm, Asbjørn Røiseland and Nils Aarsæther), *Nærdemokrati: Teori og Praksis* (2002, with Nils Aarsæther) and *Place Reinvention in the North. Dynamics and Governance Perspectives* (2007, edited with Brynhild Granås).

Gry Paulgaard   MA and PhD in social science from the University of Tromsø, Associate Professor in the Department of Education. Research on cultural identity with modernity, globalization and place enactment. Member of the 'BarentsYouth Research Network'. Among her many publications are the book *De Andre: Ungdom, Risikosoner og Marginalisering* (2003, with Kåre Heggen and Gunnar Jørgensen) and articles in *Young, Nordic Journal of Youth Research*. Member of steering committee for the Research School 'Citizenship, Encounters and Place Enactment' (CEPIN) at the University of Tromsø.

Dominic Power BA in political science and sociology from Trinity College Dublin, MA, M.Phil in politics and D.Phil in geography from University of Oxford. Associate Professor in Economic Geography, University of Uppsala. Earlier lectureships at University of Nottingham and University of Durham. Research on cultural industries, including the music, design and fashion industries. In addition to numerous international journal articles, reports and contributions, he has published *The Cultural Industries and the Production of Culture* (2004, edited with Allen Scott).

Kirsten Simonsen PhD and Dr. Phil in geography. Professor in Social Geography in the Department of Environmental, Social and Spatial Change, Roskilde University. Visiting Professor at University of Tromsø (CEPIN). Has worked in University of Copenhagen, Aarhus University and Nordplan in Stockholm. Among her many publications are *Byteori og Hverdagspraksis* (1993), *Praksis, Rum og Mobilitet* (ed. 2001), *Voices from the North, New Trends in Nordic Human Geography* (2003, edited with Jan Öhman), *Space Odysseys* (2004, edited with Jørgen Ole Bærenholdt), *Geografiens Videnskabsteori* (2004, with Frank Hansen) and *Byens Mange Ansigter* (2005).

Unnur Dís Skaptadóttir BA in anthropology from University of Massachussets and PhD from the CUNY Graduate Center, New York. Associate Professor in Anthropology at the University of Iceland. Research in economic anthropology, coping processes, migration and gender. Iceland's participant in the MOST CCPP projects since 1997. Many international publications, among others in the journals *Women's Studies International Forum* and *Sociologia Ruralis*.

John Urry Following first degrees in economics, PhD sociology from Cambridge University. Professor in Sociology and Director of the Centre of Mobility Research (CeMoRe) at Lancaster University. Visiting Professor and Doctor of Honour at Roskilde University. Editor of the journal *Mobilities*. Author of more than 30 books, among them are *The Tourist Gaze* (1990, second edition 2002), *Economies of Signs and Space* (1994, with Scott Lash), *Consuming Places* (1995), *Contested Natures* (1998, with Phil Magnachten), *Sociology beyond Societies* (2000), *Global Complexity* (2003), *Performing Tourist Places* (2004, with Jørgen Ole Bærenholdt, Michael Haldrup and Jonas Larsen), *Mobilities, Networks, Geographies* (2006, with Jonas Larsen and Kay Axhausen) and *Mobilities* (2007).

Arvid Viken MA in political science from the University of Oslo. Associate Professor at Finnmark University College. Research and very many publications on tourism development, among those are the books *Turisme, Postmoderne Kultur og Tradisjon* (2000) and *Turisme, Miljø og Utvikling* (2004) and many articles in international journals.

Anna Wojtynska First degrees from University of Warsaw. PhD student in the Department of Anthropology and Folkoristic at the University of Iceland. PhD project on migrant Polish labour in Iceland's fish industry.

Preface

The contributors to this book constitute a hybrid of approaches to 'mobility' and 'place'. Although all the contributions to this special collection of texts are in some way rooted in the North, the contributors are also establishing new routes into international debates. The book takes the form of a reader, which may be used across disciplines such as Geography, Anthropology, Sociology, Planning, Cultural Studies, Political Science and Religious Studies. It addresses undergraduate, graduate and Ph.D. students within all these disciplines, as well as international academic audiences in general.

The present work is the outcome of an exciting academic journey. It has taken as its point of departure the University of Tromsø, with its many networks and nodes. These networks are associated with rich research traditions and social theories that cut across studies of community, gender, indigenous peoples, migration, religion and tourism. This was only possible because of the Globalization from Below project (2004–2008), funded by the Research Council of Norway and awarded priority by the Social Science Faculty of the University of Tromsø. The project has been managed by the Department of Planning and Community Studies at the University of Tromsø. Interdisciplinarity and social engagement have been important qualities in this journey, which was supported by Professor Nils Aarsæther, also head of the department.

Ideas for this book have been in the making for years, but the plans really took off during the spring semester of 2006, when we, the editors, both worked in neighbouring offices at the Department of Planning and Community Studies. Based on and backed by the Globalization from Below project (uit.no/gfb), we have tried to focus and direct many of the efforts that have developed from several projects over the years: the UNESCO MOST Circumpolar Coping Processes Project; a series of NORDREGIO projects; the cross-disciplinary international doctoral Nordic Research School on Local Development (NOLD); and the cross-disciplinary doctoral school Citizenship, Encounters and Place Enactment in the North (CEPIN) at the University of Tromsø.

We thank all the contributors for their thorough work in several rounds of drafting and editing. Our workshop in Tromsø in April 2007, at which most of the contributors were present, became a central point of convergence for this process. Based on draft chapters, the workshop helped us to improve and focus the book during the last editorial round.

Besides the work of the contributors, the creation of this book has depended on numerous other forms of support. In addition to support from the Research Council of Norway, the University of Tromsø (in particular the Department of Planning and Community Studies), our co-operation and agreement with Ashgate Publishing and the professional efforts of several personnel there became crucial in what proved to be a well-organized publication process. We were further supported by the precise, careful and professional work of sociologist Eva S. Braaten, when she took over from Brynhild as Research Officer in the Department of Planning and Community Studies in 2007. Together with careful language revision by Mary Katherine Jones, this was a great help in finalizing the book.

Brynhild Granås and Jørgen Ole Bærenholdt
Tromsø and Roskilde, September 2007

Chapter 1

Places and Mobilities Beyond the Periphery

Jørgen Ole Bærenholdt and Brynhild Granås

Introduction

Mobility and place become together. While many debates tend to favour one more than the other, this book consistently investigates the intersections of mobility and place. The thematic focus is a spin-off of contemporary debates in the social sciences, explored here through research within the Northern European periphery. These approaches elucidate social transformations on the margin whilst also contributing to international academic discourses, which are otherwise more oriented towards so-called global centres. The analyses presented tell of continuous mobility engagements as part of place production, involving the remaking of places and economies, and new political and cultural projects of definition.

Images of the Northern European peripheries are varied. They may be associated with remoteness, frontier and isolation, for example, with exotic natural scenery and attractive tourist places, or with colonialism and resource extraction, for example, through traditional fishery or industrial mega-projects. This book supersedes any long-distance and from-above characterizations of this part of Northern Europe that may exist. The areas are investigated as people's lived spaces and tell of how their practices and relations enmesh places of the peripheries into a 'global' world through interdependency, connectivity and mobility.

The people concerned are multifarious, ranging from ethnic majority inhabitants, indigenous people and minorities to migrants and tourists. Some may migrate or commute for labour in old or new industries, whilst others may pursue the hybrid urban/rural lifestyle opportunities of the periphery. Our contributions demonstrate how new social scapes are enacted across distinctions such as periphery-centre, local-global, and so on. Descriptions from these positions illustrate the fact that society is 'performed through everyone's effort to define it' (Latour 1986, 275). In addition, in the cases explored in this book, societies are performed over distances and at a distance, through the social interaction, networks and fields that people enact; local communities are not *a priori* social fields (Saugestad 1996). Societies are thus made, transformed and emerge with everyday interpersonal relations, and cannot be approached as closed, holistic systems (Barth 1992). The making and remaking of periphery places involves crucial transports, connections and mobilities towards physically distant people and places.

This book demonstrates how specific relationships between mobility and place are crucial in the making of societies. The ways in which people in Northern European peripheries have been coping with distance is a powerful case of how societies are produced (Bærenholdt 2007). An integrated pursuit for this book is to analyse in greater detail how transport, communications, migration, transnationalism, tourism and travel engender social obligations and figurations (Hannam et al. 2006). As stated above, our approach to places does not represent a counterpart to this, since place is not the static and fixed contrast to mobility. Also, places are not discrete and powerless enactments: rather they are involved in the wider 'power geometries' of the processes of globalization. The social production of places therefore entails highly contested political and economic actions involved with the fundamental question of who takes responsibility for whom (Massey 2004).

Places may be enacted through their roles as arenas where people are likely to meet, or do so contingently (Bourdieu 1996). Thus, even though the establishment of a place can be planned for, places are generated through non-predictable meetings between people. These meetings may be corporeal or virtual, if not also imaginative. The present reader investigates how a wide range of practices and encounters are often also framed by images, brands and the politics of the North. These are attempts to enact places and connections, in order to sustain businesses and people's livelihoods. We aim to show how the more spontaneous encounters and the more strategic actions intersect, since they feed and frame each other.

Hence, people are 'thrown together' in particular places (Massey 2005), whilst also being involved in diverse practices of mobility (Cresswell 2006). Our approach to the spatial practices of place and mobility, in line with Massey (2005), thus focuses on interactions, multiplicity and constant processes of construction and negotiation. Places and mobilities are inherently political, but there are no universal categories of the spatial organization of societies. They are formed in particular ways through the spontaneous and political practices of people, but these specificities do not produce insular local contexts. On the contrary: the idea of local context is contested with the connectivities involved in people's practices. We are thus in a state beyond the dichotomy of the good local, so-called 'internal', control versus the bad non-local, 'external' control. Connections and encounters crucial to people's lives are often much more complex and dynamic than envisaged in such a simple dichotomy. Contexts are thus not predetermined at any scalar level, but only emerge with the practices of making and becoming places and mobilities. In the next section we explain an approach to understanding such practices.

Enactment: Invention and Emergence

Enactment combines spontaneous and political practices, and claims a crucial play across the two: On the one hand, political projects of place enactment can hardly materialize without referring to practices of connection and encounter that have already emerged. On the other hand, the emergence of connections and encounters may well have been conditioned by specific politics of place making (Nyseth and Granås 2007). Enactment also includes the term of emergence, but not solely with the connotations of spontaneous evolution (Jóhannesson 2007). The place enactment approach combines invention and emergence, and thus bypasses dichotomies between instrumental mastering and mere tactical adaptation. In parallel with the concept of coping practices (Bærenholdt 2007), it combines strategies that 'can be isolated from an "environment"' with a tactic that 'belongs to the other' (de Certeau 1984, xix).

Places are not construed out of nowhere but involve materialities, politics and imaginations, comprising people's engagement with their physical-material environment. Practices of place enactment thus directly involve nature, politics of nature and imaginations of nature (see Chapters 16, 17, 18 and 19 by Lehtinen, Benediktsson, Kraft and Guneriussen, respectively), as well as communication technologies (see Chapters 8 and 9 by Larsen and Urry and by Brekke). This book is more concerned with the critical and cultural engagements involved and less with elements of strategic place marketing (Kearns and Philo 1993; Ek and Hultman 2007). Imaginations and myths of places are used to celebrate and enchant the margins (Shields 1991). However, these practices are often ambivalent, since economic strategies, social communities and cultural meanings can easily prove a misfit for one another. Politics of place enactment are therefore likely to become messy, but this messiness can also nourish the creativity, commitment and openness crucial to change with regard to new livelihoods in Northern European peripheries (see Chapter 13 by Jóhannesson and Bærenholdt).

One way of approaching the messiness of politics of place enactment is through the double notions of politics of propinquity and politics of connectivity proposed by Ash Amin (2004). Firstly, politics of propinquity focus on the energy of practices that may follow from people being somehow contingently thrown together within a place (see Massey 2005). Such politics are 'shaped by the issues thrown up by living with diversity and sharing a common territorial space' (Amin 2004, 39); it is a coping practice 'against a presumed hierarchy of worth or order' (Amin 2004, 40). Thus, it involves enactment through the bridging and synergy among people who happen to meet and engage in a place. Secondly, enactment of place is also about a politics of relational connectivity (Amin 2004, 40) that is 'open to both local and distant actors…' (Amin 2004, 41). These are emergent politics to perform and manipulate distances (Young 2006), where enactment practises re-order relations between distance and proximity. It is not only a spatial politics but also definitively temporal, since shaping the world involves making some things present and maintaining others absent, in both time and space.

A number of instances in this book show how time and space are deeply intertwined in the making and performance of place. Simply put, since place involves presence it is both spatial and temporal (see also Bærenholdt et al. 2004, Chapter 3). This is the case, for example, with places that are more or less attached to specific resources (see Chapters 17, 16 and 12 by Benediktsson, Lehtinen and Viken, respectively): such places may come and go, along with place-specific resources. However, resources that are absent 'here and now' can also, in ambivalent ways, influence present transformations; historical continuity may frame enactment processes (see Chapter 6 by Kristiansen and Hovgaard), just as ruptures can set the course for new activities (see Chapter 15 by Granås and Nyseth). Place enactment may be about preparing for reflection through networks and creating new narratives through actions and events with global or universal appeal (see Chapter 18 by Kraft) or projects that re-enact past heroes and pathways (see Chapter 13 by Jóhannesson and Bærenholdt). Enactment policies thus play with a remixing of absence and presence, connectivity and propinquity, and the like.

Deconstructing Peripherality

This book is about the enacting of Northern European peripheries, though we are in no way trying to generalize our findings or claim validity for certain areas, such as the Nordic countries or the peripheries of the Nordic countries – not forgetting Russian parts of the Northern European periphery. More than geographical areas, it is the use of the category 'periphery' that needs to be considered here.

In the more economics-based Chapter 14 by Power and Jansson, Nordic countries are themselves considered to be peripheral, since the commodity chains under study envision a centre-periphery structure between consumers and producers. They show how manufacturing firms, which can be central job providers in the periphery, have a more contingent relation with their specific location than with their temporal market-places. Firms may move, but international trade fairs are indispensable and have become nodes in global networks.

While the asymmetries embedded in such relations may be valid for some business relations, this book contributes primarily with more critical considerations of the notion of the periphery and the centre-periphery dichotomy. In Chapter 5, Gry Paulgaard argues in favour of transcending dichotomies such as centre-periphery, especially when associated with other binaries such as future-past. Paulgaard shows how young people, when negotiating their identifications, still have to cope with such binaries and even hierarchical orders, despite the tendency of academics to view centre-periphery models as 'outdated'. A discursive backcloth for young people's negotiations is Norwegian intellectuals' more or less romantic identity politics, which have praised small towns and villages, as well as associating the periphery with nationalistic projects (see Bærenholdt 2007). Eager to be modern, youngsters produce hierarchical orders among localities within the periphery as measures of 'degree of modernity'. Such centre-periphery narratives are laden with power, as they have also become part of how places are practised (see Chapter 2 by Simonsen). However, there are also approaches to place – focusing on a more direct sensing of and being along places – that do not address or value the meaning of such narratives and identity politics (such as Chapter 3 by Mazullo and Ingold).

The centre-periphery dichotomy needs to be deconstructed. Hence it is a question of the extent to which social scientists should use the kind of relational centre-periphery model applied in economic geography (see Chapter 14 by Power and Jansson). However, when youngsters in the far North engage in their world with concepts not far removed from those of economic geography, there are obvious limits to the project of deconstructing peripherality. Meanwhile, several chapters in this book suggest a re-centring of our awareness of the 'radius' or 'concentric side of human community building' (see Chapters 5 and 16 by Paulgaard and Lehtinen, respectively), by focusing on how people live and make their lives meaningful, in spite of attempts to displace them. Ironically, such endeavours as forms of resistance to outside dominance may re-affirm the dichotomies questioned.

When peripheries tend to re-emerge, this is not only because certain projects of re-centring produce new 'internal' peripheries of those not included (see Chapter 6 by Kristiansen and Hovgaard). It is also because the periphery is being endlessly reproduced as a part of personal or political identity projects. This may occur in the form of individual searches for meaning and authenticity in life (see Chapter 4 by Birkeland). It may also occur as part of a broader pattern of travelling, drawing on the image of the North or even nordicity/nordism; some are in search of a frontier, where acknowledgement may depend on one's capabilities in outdoor life (see Chapter 12 by Viken) and an understanding of the 'rules of the North to survive in the backwoods', as a way of 'measuring one's "nordicity"' (see Chapter 16 by Lehtinen).

The North is associated with an image of the suppressed in need of new stories, beyond primitivism, smallness, periphery, nature and tradition (see Chapter 18 by Kraft). Paradoxically, such projects – like the Tromsø campaign to host the Winter Olympics – tend to re-invest the North with wilderness and the 'magic' of the savage (see Chapter 19 by Guneriussen). However, there are definitely distinctions in landscape views: Icelanders are proud of 'their' nature (see Chapter 17 by Benediktsson), while Polish labour migrants may perceive the landscapes of Iceland as obstacles and as less authentic than those where they came from (see Chapter 10 by Skaptadóttir and Wojtynska). The work-centred approach to places in Iceland that is displayed by labour migrants differs from the outdoors-centred, playful engagement with nature among the temporary inhabitants of Svalbard (see Chapter 12 by Viken).

Ari Lehtinen shows how place-making and community-building in a trans-border area combine memories of relational displacement and belongingness in 'concentric initiatives of imaginary relocation'. He suggests a critical place enactment approach to forest conservation in the Greenbelt that involves shared memories as well as identity politics (thus in line with suggestions made in Chapters 2 and 5 by Simonsen and Paulgaard, respectively). This approach contrasts the highlighting of centre-periphery within place marketing and economic geography with the shared (concentric) memories of people living there. Karl Benediktsson similarly suggests approaching conservation as a politics of nature, involving stories of construction workers, imaginations of protest and mobile protectors to the mega-project of hydropower construction in Iceland. Where mountains are removed and valleys are flooded, academics' 'constructivist play' with nature needs to engage with how people feel about such irreversible changes. 'Even if a view of place as a relational enactment is adopted, place must still be recognized as an enormously important locus of affect and emotion' (Chapter 17 by Benediktsson).

Peripheries can thus clearly be deconstructed and dichotomies superseded. Also, as we shall see in the next section, there are connections and encounters going on that achieve this in practice. However, such mobilities, events and cross-cutting connections do not transcend the idea of place as something of meaning and value. On the contrary, as we have seen above, mobilities of various kinds actually contribute to accentuating the meaning of places.

Place and Mobility Practices

The general approaches in this book are via mobility and place, whilst our intention is to identify and explore the specific variations of places and mobilities. Both places and mobilities are multidimensional. Places are thus material, social and cultural, and include various types of practices. Mobilities are about the mobility of people, as well as things and information, and these various mobilities are intertwined in different ways.

Importantly, mobilities such as tourism, migration and commuting intersect, since they feed into and produce one another (Williams and Hall 2006). For example, tourist visits and tourist performances sometimes become the first step to moving to another place; this may consequently imply a new commuting practice. Additionally, more people tend to establish 'second homes' which become more than just a holiday place, but a place for work, family and recreation in large slots of the annual time budget. In Chapter 7, Siri Gerrard tells of the mobile practising of place in a small fishing village: holiday residents, Sámi pastoralists and tourist crowds on their way to visit the North Cape make the place, together with local inhabitants who stay for part of the year in holiday homes in the North or the Mediterranean. Such mobile practices contribute to making the remote fishing village well-connected, becoming a place of conjunctions (Gerrard, with reference to Chapter 2 by Simonsen). Siri Gerrard's chapter exemplifies how kinds of mobilities involved, and their motives, are multiple and thus display different meanings of place.

As Kirsten Simonsen elaborates in Chapter 2, the intertwining of mobility and place includes power relations and stories of travel, and place becomes a 'locus of encounters, the outcome of multiple becomings' (Chapter 2). Though places are always in the making, they are simultaneously embodied emotions, narratives and memories. Such emotions, narratives and memories sometimes become the locus of festivals, where places then are made into events which enable people to meet up and do something together (Massey, 2005, see Chapter 7 by Gerrard). Stability of place is thereby underpinned by mobile practices of travel.

Massey's (2005) concept of 'place as throwntogetherness of people' encircles the time-specific and social-relational characteristics of place. We suggest a consideration of how the concept of throwntogetherness can also be used when exploring the mobile paths of human activities. In Chapter 3, Nuccio Mazullo and Tim Ingold suggest a mobile phenomenology of 'being along' where 'places can only occur along paths of movement…' Their ethnography, centring on Northern Finland, suggests an understanding of place along the paths the Sámi people follow, summed up by the concept of 'metakinesis(scape)'. Mazullo and Ingold are less concerned with the sociability of multiple people thrown together along routes. However, their conceptualizations are interestingly open to approaching the time-specific articulation of social life through a closer engagement with movement in the material landscape.

In line with the tradition of humanistic geography, one might state that personal identity and belonging to a place can become stronger through mobility. While placelessness is here conceived as a concept for the pathological, this is not because of mobility, since mobility is a way of finding meaning and ways to places and belonging (see Chapter 4 by Birkeland). For Birkeland, place becomes an explicit critique of modernity, since she is occupied with how humans sense the authentic. Birkeland argues that we should leave behind the notion of 'having place' and make room for a politics of place concerned with the making of or being a place, for example a community that is meaningful, whilst also being both human and non-human.

The social relational approach to place, inspired by Massey, occupies a strong position throughout this book. The approach opens up place – but also mobility – as concepts for societies, social encounter and cultural multiplicity. Even though Mazullo and Ingold, for example, and also Birkeland place less emphasis on this aspect, several chapters make use of a social relational approach to place, particularly as elaborated in Kirsten Simonsen's opening chapter. While Simonsen is less occupied with the material and with technology, she is more concerned with the social and with narratives, stories and connection with others. This approach shares a relational understanding of practices with other chapters and approaches, such as those by Larsen and Urry (Chapter 8) and Jóhannesson and Bærenholdt (Chapter 13), and the more empirical chapters in this book.

In Chapter 6, Kristiansen and Hovgaard analyse the increasing mobilities of people within the Faroes. This chapter, together with that of Larsen and Urry (Chapter 8), argues convincingly against Putnam's (2000) perception of mobility and urban sprawl as a threat to social capital and social cohesion: as an alternative, they suggest that mobility and network capital condition social capital (see also Urry 2000). Another chapter in line with those mentioned above is Arvid Viken's study of Longyearbyen in Svalbard (Chapter 12). Viken shows the role of long-distance networks in a place of extreme, but also institutionalized, mobility. He argues against Putnam and suggests that these mobilities stabilize society, social relations and social capital, rather than rendering society and place on Svalbard more fragile.

Networking and mobile technologies do not mean the end of places or distances: they only contribute to material, social and cultural reconfigurations of places and distances. Social networks may be practised with the help of technologies, such as the Internet and mobile phones (Urry 2000; see also Chapters 8 and 9 by Larsen and Urry, and by Brekke). Larsen and Urry argue that social sciences have over-focused on propinquitous communities, equating 'near' with 'close and meaning full'. However Simmel, writing about the stranger in the city, has already shown how distant people can be from each other in proximity (see Allen 2000). Furthermore, Larsen and Urry show that ideas and concepts from Castells (1996) about 'space of flows' are overly cognitivist. Distance still matters a lot to people, as does place.

The complex intersections of place and mobility practices are particularly illuminated through migration studies. Firstly, in her chapter about young refugees (Chapter 9), Marianne Brekke presents examples of transnationalism as long-distance, cross-border connections. Young refugees perceive and practise the town in which they live as a transit place; internet and mobile phone connections with people in other places are central within this practice. Meanwhile, the local library becomes a meeting-place for refugees to connect with other places. Place and mobility practices of refugees obviously involve power geometries. Secondly, Chapter 11 by Flemmen and Lotherington shows how this is also the case with transnational marriages between Norwegian men and Russian women. Their variety of case studies reveals relations that are more complex than the overall asymmetry between 'Western' men and 'non-Western' women. The diversity of transnational marriages plays on different kinds and amounts of network capital and the cultural logics of desire, tied into a wider political economy that people cope with by pragmatic arrangement. This involves various practices of place and mobility, such as dual citizenship, two-home households and the like. The practices of transnational married couples are examples of why we need to reconceptualize our understanding of society. Thirdly, Chapter 10 by Skaptadóttir and Wojtynska argues for a shift in attention towards mobility and bifocalization. The authors show how Polish labour migrants to remote fishing villages in Iceland live in a state of movement 'within family networks'. The fishing village in Iceland becomes a meeting place and a postnational zone. Although distance matters and migrants do not go to Poland as often as they would like, they are in fact living a bifocal life, simultaneously living in two places and betwixt these two places.

The social sciences need to acknowledge the ambivalence and contradictions involved in these ways of life as something that has always been present and has now taken other forms. We hope that this book contributes to this. Being along paths and routes of travel, between and across significant places, mobility and place come together in people's practices. Though now in new ways, people have for a long time used various technologies of mobility of things and information to keep in contact. Place and mobility practices may thereby reconfigure nearness and distance, propinquity and connectivity. However, distance still matters, not the least in the ways that people enact social connections.

Outline of the Book

The chapters that follow are organized into three sections. The first section, 'Placing Mobility', takes off with three chapters where place and mobility are conceptually interpreted, based on practices and people's sensing of place. These are followed by three chapters exemplifying and exploring such practices and sensing of place within localities in the Faroes and the far north of Norway.

The second group of chapters, under the heading of 'Connections and Encounters', addresses the mobile and networking practices of people, using mobile phone and the Internet to keep in contact – thereby also stabilizing not only societies at a distance but also the more localized places of encounters in the North. Whilst all five chapters address networking across distance, both theoretically and empirically, they look especially at transnational practices among migrants, refugees, transnational married couples and temporal workers.

While the first section of chapters may be more focused on places, and the second section more on mobilities, the chapters in the last section, 'Mobilizing Place', try in various ways to tie these together, but with a more specific focus on the politics of enacting places in Northern Europe. The first three chapters address various economies and policies, combining propinquity and connectivity in the case of tourism, cultural industries such as furniture production, and the transport of iron ore and other goods. These are followed by two chapters which look critically at the politics of nature involving place enactment, between nature conservation and new concrete constructions, such as the hydropower mega-project in Iceland. This leads to the two final chapters, which cast light on the events and images used in attempts to re-enact the very particular northern city of Tromsø – which, by the way, has also been the meeting-place for the authors involved in this book, attracting their mobilities. We invite readers to join us.

References

Allen, J. (2000), *On Georg Simmel: Proximity, Distance and Movement*, in Crang and Thrift (eds).

Altern, I. (ed.) (1996), *Lokalsamfunn og lokalsamfunnsforskning i endring* (Universitetet i Tromsø: Institutt for Samfunnsvitenskap).

Amin, A. (2004), 'Regions Unbound: Towards a New Politics of Place', *Geografiska Annaler* 86B:1, 33–44.

Barth, F. (1992), 'Towards Greater Naturalism in Conceptualizing Societies', in Kuper (ed.).

Bærenholdt, J.O. (2007), *Coping with Distances: Producing Nordic Atlantic Societies* (Oxford: Berghahn Books).

Bærenholdt, J.O., Haldrup, M., Larsen, J. and Urry, J. (2004), *Performing Tourist Places* (Aldershot: Ashgate).

Bourdieu, P. (1996), 'Et steds betydning', in Bourdieu, *Symbolsk makt – Artikler i utvalg* (Oslo: Pax).

Castells, M. (1996), *The Rise of the Network Society, The Information Age* Vol. 1 (Oxford: Blackwell).

de Certeau, M. (1984), *The Practices of Everyday Life* (Berkeley: University of California Press).

Crang, M. and Thrift, N. (eds) (2000), *Thinking Space* (London: Routledge).

Cresswell, T. (2006), *On the Move: Mobility in the Modern Western World* (New York: Routledge).

Ek, R. and Hultman, J. (eds) (2007), *Plats som produkt* (Lund: Studentlitteratur).

Hall, C.M. and Williams, A.M. (eds) (2002), *Tourism and Migration, New Relationships between Production and Reproduction* (Dordrecht: Kluwer).

Hannam, K., Sheller, M. and Urry, J. (2006), 'Editorial: Mobilities, Immobilities and Moorings', *Mobilities*, 1: 1, 1–22.

Jóhannesson, G.T. (2007), *Emergent Tourism: An Actor-Network Approach to Tourism Economies* (PhD. thesis, Department of Environmental, Social and Spatial Change, Roskilde University).

Kearns, G. and Philo, C. (eds) (1993), *Selling Place: The City as Cultural Capital, Past and Present* (Oxford: Pergamon Press).

Kuper, A. (ed.) (1992), *Conceptualizing Society* (London: Routledge).

Law, J. (ed.) (1986), *Power, Action and Belief* (London: Routledge and Kegan Paul).

Latour, B. (1986), 'The Power of Association', in Law (ed.).

Massey, D. (2004), 'Geographies of Responsibility', *Geografiska Annaler* 86B:1, 5–18.

Massey, D. (2005), *For Space* (London: Sage).

Nyseth, T. and Granås, B. (eds) (2007), *Place Reinvention in the North. Dynamics and Governance Perspectives* (Stockholm: Nordregio).

Putnam, R.D. (2000), *Bowling Alone* (New York: Simon & Schuster).

Saugestad, S. (1996), 'Mellom modeller og virkelighet – Lokalsamfunnsforskning i Tromsø, Norge og verden', in Altern (ed.).

Shields, R. (1991), *Places on the Margin* (London: Routledge).

Urry, J. (2000), *Sociology Beyond Societies: Mobilities for the Twenty-first Century* (London: Routledge).

Williams, A.M. and Hall, C.M. (2002), 'Tourism, Migration, Circulation and Mobility: The Contingencies of Time and Place', in Hall and Williams (eds).

Young, N. (2006), 'Distance as a Hybrid Actor in Rural Economies', *Journal of Rural Studies* 22: 253–66.

PART 1
Placing Mobility

Chapter 2

Place as Encounters:
Practice, Conjunction and Co-existence

Kirsten Simonsen

Introduction: The End of Place?

Mobility and place, the most significant concepts of this book, are in many contemporary accounts represented as opposites. The way in which places are tied into global flows of people, significations and things, it is argued, is changing our apprehension of space and time. A combination of information technologies, increased mobility and consumer society has been blamed for an accelerating erosion of place. More and more of our lives, it is said, take place in environments that could be anywhere – that look, feel, sound, and smell the same wherever in the world we might be. Fast food outlets, shopping malls, airports, high street shops, and hotels are more or less the same wherever we go. They are localities symbolically detached from the local environment and therefore contributory to a thinning out of the meaning that ascribes significance to place (see also Cresswell 2004).

These accounts of the erosion of place appear in two different forms. The first of these have deep roots in critiques of modernity and draw basically on a figure of *loss*. A classic text in that group comes form the humanistic geographer Relph's account of *Place and Placelessness* (1976), which was written long before current narratives of 'globalization' and 'circulating entities'. Relph formulates his critique through the opposite (Heideggerian) notions of 'authenticity' and 'inauthenticity' and makes a direct connection between mobility and inauthentic placelessness. Various forms of increased mobility together with 'mass culture' dilute authentic relations to place, and places become 'other directed' and more alike across a globe of transient connections. In a similar way, Augé (1995) argues that the facts of 'supermodernity' render necessary a radical rethinking of place. His notion of 'non-places' refers to sites marked by their transience – the preponderance of mobility. Augé's notion of 'non-places' does not have the same connotation of loss as Relph's 'placelessness'. It rather calls for a rethinking of culture consistent with the fleeting and ephemeral character of 'hypermodern non-places' such as freeways, airports, supermarkets and so on. This comes closer to what can be called a *celebration* and 'ontologization' of ceaseless mobility, for example strongly formulated by Thrift (1996):

> What is *place* in this new 'in-between' world? The short answer is – compromised: permanently in a state of enunciation, between addresses, always deferred. Places are 'stages of intensity', traces of movement, speed and circulation. One might read

this depiction of 'almost places'…in Beaudrillardian terms as a world of third order simulacra, where encroaching pseudo-places have finally advanced to eliminate places altogether. (1996, 289)

The very writing of this chapter suggests that I am sceptical to the extreme view of the relationship between mobility and place in different ways propagated by Relph, Augé and Thrift. Of course mobility affects place (and vice versa) but the relationship between them should be seen, not as one of erosion, but as a complex intertwining contributing to the construction of both. In the following, I (like many others) shall pursue a conception of place that comply with this view, but first a few words on major positions in the theorization of place.

The Betweenness of Place

This heading is adopted from a book by Entrikin (1991) bearing it in the title. It refers to the polarity existing in both everyday life and academic discourse between two aspects of our understanding of place: a relatively objective, naturalistic conception of place and a relatively subjective, existential sense of place. This polarity comes into existence through binary oppositions such as naturalism/existentialism, objective/ subjective, explanation/understanding, and science/art. The tendency has been to reduce one side to the other, by suggesting that all that is real from the subjective view is reducible to the objective and vice versa.

The naturalistic line has supported a relatively descriptive approach to place. Emphasis has been on materiality and 'nature', and places are conceived of as landscapes synthesizing the material and the cultural. Morphology and landscape become organic wholes, and often scientific support has been sought in biological sciences. Traditional analyses of regions and places, as for example the American cultural geographer Carl Sauer (see Entrikin 1984), used evolutionary biology and natural history as models for the cultural history of place, or they grounded their work in a sense of terrestrial unity and a naturalistic conception of social science. Later versions rely on ecological holism based on the metaphor of organism or on the functional holism of system analysis.

The existential line covers humanistic approaches to place giving emphasis to experience and meaning. They start from a first person perspective, not because of a primary interest in consciousness and subjectivity, but because objects or material things for them are 'intended objects', coming into existence through action, perception and intended use (see for example Casey 1993). From this phenomenological perspective, places are constituted by the way in which we interact with them – by experiences and meanings created in this interaction. They are the spatial dimension of human 'lifeworlds', the pre-reflexive world of experience and familiarity fundamental in any ideas of everyday life. These accounts have given rise to a range of concepts connected to experience and meaning – such as 'sense of place', 'place identity', 'personality of place', or 'spirit of place' (*genius loci*) – now part of nearly any vocabulary of place.

Notwithstanding differences and opposites one aspect of place seems to be present in nearly any account; it is the one of *specificity* or related notions of uniqueness, the concrete and the ideographic. That is also the case in approaches importing a stronger element of social theory. They basically approach places through notions of practice, process and change, and their main arguments can be summarized in two general statements. The first one concerns the 'social construction of place' and underscores places as instances of more general social practices and processes (for example Pred 1984; Paasi 1986, 2002). Places are never finished but always becoming – they are products of social relations and interactions, which are continuously constructed, laid down, decayed and reconstructed. The other (related) statement is that 'place matters' (see for example Massey and Allen 1984). That one is based on the simple fact that social processes take place and places vary. Social forces shape places and are in turn shaped by places. In other words, when places are once constructed, they influence the way in which social processes and activities are performed.

Continuing from these two statements, in the following I shall attempt to take the concept of place further in a manner that rests on a theory of practice and takes into account the significance of mobile bodies.

Practice and the Conjuncture of Place

The conception of place pursued here starts from a social ontology of practice, that is, an account of social life maintaining that human lives hang together through a mesh of interlocked practices, as a constitutive part of which this hanging together occurs (see among others Simonsen 2003). That means that practices constitute our sense of the world, and that subjectivity and meaning are created in and through practice. One inspiration for that has been Heidegger's existential phenomenology – in particular as formulated in *Being and Time* (1962). In this work, 'Dasein' or human 'being-in-the-world' is described as an existential 'facticity' – as a practical, directional, everyday involvement. Our concern with the environment, it argues, takes form by way of tools and articles for everyday use as well as useful products and projects – all together designated as 'equipment' (Zeug). 'Being-in-the-world' is the everyday skilful coping or engagement with an environment including things as well as other human beings. That means that our 'environment' does not arrange itself as something given in advance but as a totality of equipment dealt with in practice. Heidegger describes very simple skills – hammering, walking into a room, using turn signals, and so on – and shows how these everyday coping skills involve a familiarity with the world that enables us to make sense of things and to find our way about in our public environment. He thus demonstrates that the only ground we have or need to have for the intelligibility of thought and action is in the everyday practices themselves, not in some hidden process of thinking or of history (Dreyfus and Hall 1992).

An important characteristic of lived, everyday practices, which is not explicitly acknowledged by Heidegger, is that they are intrinsically corporeal. In order to accomplish that we can turn to Merleau-Ponty's sensuous phenomenology of lived experience, which identifies the body as part of a pre-discursive social realm based on perception, practice and bodily movement (Merleau-Ponty 1962). Lived experience, to him, is always and necessarily embodied, located in the 'mid-point' between mind and body, or between subject and object – an intersubjective space of perception and the body. In it, perception is based on practice; that is, on looking, listening and touching and so on as acquired, cultural, habit-based forms of conduct. Perception, from this perspective, is not an inner representation of the outer world, but rather a practical bodily involvement. It is an active process relating to our ongoing projects and practices, and it concerns the whole sensing body. This means that the human body takes up a dual role as both the vehicle of perception and the object perceived, as a body-in-the-world – *a lived body* – which 'knows' itself by virtue of its active relation to this world. The 'body-subjects' are not locked into their private world, but are in a world that is shared with 'others' – a genuine interworld of humans and nonhumans.

However, the corporeality of social practices does not only concern the sensuous, generative and creative nature of lived experiences, but also how these embodied experiences themselves form a basis for social action. Bourdieu (1977, 1990), for instance, talks about 'habitus' as embodied history which is internalized as a second nature. Social structures and cultural schemes are *incorporated* in the agents and thus function as generative dispositions behind their schemes of action. The result of this socialization of practice is that social orders are established in practice. The arrangements of people, artefacts, organisms, and things that form the site of the social are laid down primarily in the interweaving and interrelated nexuses of practices that entities of the first of these sorts carry on (Schatzki 2002). The endless movement and becoming of the social are constituted in intersecting waves of doing.

Thinking about place from this point of view rules out any ontologization of the concept. Places are never givens or essentials. The theoretical starting point is space or rather the *spatiality* (and temporality) of practices. As a first approximation, then, a place can be seen as a *specific articulation* of social practices, social relations and materiality as well as experiences, narratives and symbolic meanings of the place held by its different users.

It is important to maintain that such a conception of place is a highly dynamic one (see also Massey 1994). If place is a specific conjunction of social practices and social relations, it will consist of particular interweavings or networks which have over time been constructed, consolidated, decayed and renewed. Places are not by definition bounded areas with fixed boundaries (even if boundary construction does occur) but specific, unique moments of social relations and social experiences, where some of these relations and experiences are constituted within what we happen to define for that moment as the place itself while others are based on far larger scales and connect the place to other places. This is the background from which Doreen Massey talks about the possibility of a 'global sense of place' (1994, 146) as a description of the way in which the global is present in the local, and the place is constituted in interaction between both of them.

The reason why the emphasis on the open and dynamic character of place is so important is the implicit meaning, which has often been ascribed to the concept in both academic and planning discourses. These discourses almost construe place as absolute space and signify it (implicitly or explicitly) with notions such as 'community', 'home' or 'stasis'. In these formulations, place becomes an essentialist concept, characterized by closure and timeless stasis. These imaginations are based on a range of binary dichotomies locating place on the static and passive side; they are amongst others time/place, flows/places, movements/stasis, and global/

local. This kind of thinking on place carries two problems, identified before (see for example Rose 1993; Massey 1994; Simonsen 1995) but still in work. The first ones are associated with discussion of the meaning of place in the (post)modern world (for example Harvey 1989, 1996; Laclau 1990). In these discussions, place becomes a refuge or a delimitation against insecure surroundings. It represents a nostalgic search for roots or identity up against the motility and instability of the (post)modern world. Even if these tendencies do exist, the generalization of them carries unfortunate political overtones. Place and sense of place are looked at as escapism from the dynamics and the changeability of 'real life' that have to be overcome. While time is aligned with movement and progress and a progressive project of 'becoming', place becomes stasis and reaction – a passive 'being'. The second problem arises when the content of the dichotomies is taken into account. With time, for instance, are aligned ideas of progress, civilization, science, politics and rationality, while place in this connection is aligned with stasis, reproduction, emotion, aesthetics and the body. It is not difficult to see that this system of dualisms is equivalent to the ones that we often find connected with masculine and feminine – or in other words that place ends up by being coded feminine. This coding is in accordance with the ways of thinking where place is conceptualized in notions of 'community', 'dwelling' or 'home' – a nostalgic and idealized image of home, which in fact is connected to a naturalization of traditional gender roles and a masculine dream of the idealized 'Mother'.

The alternative concept of place offered here as the first approximation sets up two major dimensions. In the first one, I follow Massey (1994, 2005) and emphasize a dimension of *specificity*: it emerges through the above characterization of place as a specific articulation of different social practices, narratives, relations and materialities. In this conception, place also turns out to be a concept of difference. Places become different, unique conjunctions of networks of practices and relations, not only a set of static conditions that the dynamic processes effect, but co-producing constellations in continuous construction and change.

The second dimension is an *existential-symbolic* dimension referring to the symbolic signification of place and its meaning to inhabitants and users. This dimension includes our ascription of symbolic meaning to places and their possible involvement in our formation of identity, but also – to paraphrase Merleau-Ponty and Bourdieu – our 'sense of place': our practical knowledge of it, familiarity with it, and ability to get about in it. It also connects to what Lefebvre (1991) designates as 'lived space' and the ability of different people to appropriate space. It is important to underline that this dimension is neither a passive nor a static one. It can be elaborated from the point of view of Heidegger's existential phenomenology, but from another reading of it than the one that focuses on 'dwelling' and see that as synonymous with home, authenticity and nostalgia. The employed interpretation builds on Heidegger's understanding of being-in-the-world as an existential facticity; a practical involvement and relation to the environment in the form of humans and nonhumans. Significations and meanings developed in such relations are not static or passive but under continuous negotiation and alteration. In this understanding, therefore, existence or 'being' cannot be static, and it therefore renders dubious the above-mentioned opposition between 'being' and 'becoming' and its reference to place and time respectively.

The emphasis on the dynamic aspect, which is then an integrated part of the dimension of specificity as well as the one of existence and signification, indicates that what primarily is on the agenda in relation to the conception of place in question is, not so much place in itself, but rather the constitution of place. It is the continuous construction and reconstruction of places in all its complexity and indeterminacy, involving the conjunction of institutional, corporeal, symbolic and material elements.

Constitutive Operations

The practice approach sets focus upon spatial operations by which humans produce the specificity and symbolic signification of places through their doings on and with the surroundings. Through specific skills, sensibilities and dispositions they manipulate 'objects' (and other 'subjects') and ascribe meanings to them. In this section I shall try to elaborate the conception of place by specifying four different 'ways of operating' in the construction of place.

Embodiment

The first operational scheme grows out of the corporeality of practice – it is about moving bodies that are folded into the world by virtue of senses and gestures and about the spatiality of these bodies. To explore that we can initially turn to Merleau-Ponty and his distinction between the spatiality of things and the spatiality of human bodies as being spatialities of *position* and *situation* respectively. This goes for temporality as well, and it means that we should avoid thinking of our bodies as being *in* space or *in* time – they *inhabit* space and time:

> I am not in space and time, nor do I conceive space and time; I belong to them, my body combines with them and includes them. The scope of this inclusion is the measure of that of my existence. (Merleau-Ponty 1962, 140)

This means that active bodies, using their acquired schemes and habits, position their world around themselves and constitute that world as 'ready-to-hand', to use Heidegger's expression. And these are moving bodies 'measuring' space and time in their active construction of a meaningful world.

Lefebvre (1991) adds to that with his stronger emphasis on materiality and the production of space. He establishes a material basis for the production of space consisting of *the spatial body* that both *is* and *has* its space; it produces itself in space at the same time as it produces that space. It is

> ...a practical and fleshy body conceived of as totality complete with spatial qualities (symmetries, asymmetries) and energetic properties (discharges, economies, waste) (1991, 61)...A body so conceived, as produced and as the production of space, is immediately subject to the determinants of that space...the spatial body's material character derives from space, from the energy that is deployed and put to use there. (1991, 195)

Like Merleau-Ponty, Lefebvre assigns an important role to the body in the 'lived experience'. As a part of that, the body constitutes a practico-sensory realm in which space is perceived through sight, smells, tastes, touch and hearing. It produces a space which is both biomorphic and anthropological. The relationship to the environment is conducted through a double process of orientation and demarcation – practical as well as symbolic.

When Lefebvre refers to the energy of the body, it is not only a material/ biological notion. With reference to Nietzsche, he emphasizes the Dionysian side of existence according to which play, struggle, art, festival, sexuality and love – in short, Eros – are themselves necessities. They are parts of the transgressive energies of the body. Further, it is important to notice that this concerns not only the material and meaningful production of space, but also the capacity to transgress the 'everydayness' of modern life. It involves participation and appropriation of space for creative, generative, bodily practices, as formulated for instance in the 'right to the city' (Lefebvre 1996).

This means that places are constructed through what Merleau-Ponty (1968) would call the 'intertwining' between 'the flesh of bodies and the flesh of the world' involving the situatedness of the body and its active production of space.

Emotion

Following from embodiment, the second type of constitutive operations is emotion. From a phenomenological point of view, we are never 'un-touched' of the world around us; that is what Heidegger's notion of *moods* or *Stimmung* implies (Heidegger 1962; Guignon 2003). Moods are basic human attributes, but they are not inner physical and psychic conditions. We should rather see them as an attunement – a contextual significance of the world, associated with practices, lifemode and social situation. The same ideas of situatedness and of the collapse of the distinction between 'inner' and 'outer' are involved in Merleau-Ponty's visions on emotion. Emotions are *situated corporeal attitudes*, ways of being and acting in relation to the world (Merleau-Ponty 1962; Crossley 1996). They are inseparable from other aspects of subjectivity, such as perception, speech/talk, gestures, practices and interpretations of the surrounding world, and they primordially function at the pre-reflexive level. Emotion, then, is a way of relating. It is a form of *living meaning* which is communicated and 'blindly' apprehended through corporeal intentions and gestures that reciprocally link one body to another. Emotions in this way are part of the 'system' that body-subjects form with others. They are intersubjectively or intercorporeally constituted, shaping and being shaped by relations between body-subjects, and thereby basically public and *relational*.

Thus, everyday encounters with other fleshy and sensuous bodies affect us, and this affection can take the form of different emotions such as love, desire, hate or fear. So the sense of mutuality involved in the phenomenological account should not be mistaken for harmony. Nevertheless, a 'pure' phenomenological approach remains limited insofar as it does not appreciate differences among bodies and power relations involved in intercorporeal meetings. Power relations occur in connection with deviations from the 'neutral' body – such as sex/gender, skin colour, age, disability and sexuality. It challenges Merleau-Ponty's idea of the social body as a body opening up into the fleshy world of other bodies. For this world is a differentiated world, and in such a world what is meant by the social body is more often than not 'precisely the effect of being with some others over other others' (Ahmed 2000, 49). The social body is also an imaginary body that is created through the relations between bodies recognizable as friendly and/or strange. Familiar bodies can be incorporated through a sense of community, being with each-other as like bodies, while strange bodies more likely are expelled from bodily space and moved apart as different bodies. In this way 'like' bodies and 'different' bodies do not just precede the bodily encounters of incorporation or expulsion, likeness and difference are also produced through these encounters. What Sara Ahmed calls 'strange encounters' are not only visual, but also tactile: just as some bodies are seen and recognized as stranger than others, so too are some skins touched as stranger than other skins. This of course also involves 'situated corporeal attitudes' or emotions. Various familial relations involves particular forms of emotion and ways of touch, while the recognition of some-body as a stranger – a body that is 'out-of-place' because it has come too close – might involve a fear of touching.

Emotional operations, then, can be the way in which places live and are inscribed into our bodies in the form of general attunement, attachment or familiarity, but they also constitute place through attitudes to the other in the form of enthusiasm and joy or border anxiety, distancing, and construction of boundaries.

Narrative

Connected to embodied practices place is also constructed through narratives. Narratives can be thought of as spatio-temporal operations connecting future, past and present – or rather the mental acts of expectation, memory and attention (Riceour, 1984). The contribution from de Certeau (1984) in a straightforward way illustrates the spatiality of narratives (see also Simonsen 2004). He emphasizes the relationship between practices and narratives as well as the spatiality of both and perceives narratives as spatial trajectories:

> Every story is a travel story – a spatial practice. For this reason, spatial practices concern everyday tactics, are part of them, from the alphabet of spatial indication ('it's to the right', 'take a left'), the beginning of a story the rest of which is written by footsteps, to the daily 'news' ('guess who I met at the bakery?'), television news reports ('Teheran: Khomeini is becoming increasingly isolated'), legends (Cinderellas living in hovels), and stories that are told (memories and fiction of foreign lands or more or less distant times in the past). These narrated adventures, simultaneously producing geographies of actions and drifting into the commonplaces of an order, do not merely constitute a 'supplement' to pedestrian enunciation and rhetorics. They are not satisfied with displacing the latter and transposing them into the field of language. In reality, they organize walks. They make the journey, before or during the time the feet perform it. (1984, 115–16)

What de Certeau is arguing in this comment is that narrative is conditioning a map – a 'cognitive map' (cf. Jameson 1991) – but one that is less inclined to produce an overall vision than to represent everyday life. It is composed by a range of active narrative operations or speech acts and always related to spatial practices at different scales. Stories, in de Certeau's words, are transforming 'places into spaces and spaces into places' (1984, 118) – as spatial trajectories, they every day traverse and organize places, for instance by making them habitable and meaningful, or by selecting them and linking them together in interscalar relations.

Narratives then create storied pathways to live by, including performance conventions, plots, order, myths and so on. Exploring such narrative resources for creating and experiencing our surroundings, therefore, can institute an understanding of the place as a collection of stories. Narratives make places habitable and believable, they recall or suggest phantoms and they organize the invisible meanings of the places. The temporality and spatiality of narrative then establish how memories and expectations intersect in present dispositions for action and the constitution of spatial imageries; how narrative figures such as tours, maps, boundaries and bridges are integral parts of the construction of social life and, in particular, the organization of spaces and places; and how narrative power strategies, involved in processes of territoriality and othering, constitute hierarchies of places at different scales – from the body to the global. Conflicting narratives, then, are an important part of the multiscalar construction of place.

Memory

As indicated above, the construction of place through embodied practices and narratives also works through the spatialization of time – or memory. A feasible way of approaching memory might be through phenomenology of time. Time is here considered constitutive of human being-in-the-world and, particularly important here; it is seen as a structural connection of future, past and present (Heidegger 1962, Ricoeur 1984). These three dimensions are further seen as moments of mental action – as acts of expectation, memory and attention, respectively – each considered not in isolation but in interaction with one another. From that point of view, memories are organized and called up by attention to the present and the future or, in other words, by being-towards them. This is of course not only an individual activity, and many scholars have employed Halbwachs' seminal work on *collective memory* (1992/1951) to explore it as a social activity expressing and holding together group identities.

The cultural practice of memory is often a contested one, more or less connected to 'invented traditions' (cf. Hobsbawn and Ranger 1983), and object of social and political conflict. But one thing that (conflicting) groups share in their efforts to appropriate the past is an embedding of memories in place as a 'site of memory'. This activity often takes a material and embodied form (see also Hoelscher and Alderman 2004). First of all, the constitutive relationship between memory and place is obvious within the realm of *material culture*. Landscapes, monuments, memorials and museums are sites suitable to impart certain elements of the past and (subsequently) to forget others. Secondly, the relationship can be *performative*. The whole set up of performances such as rituals, festivals, pageants and civic ceremonies serve as major ways in which (imagined) communities carry out the cultural practice of memory (see for example Thrift and Dewsbury 2000). The contested relationship between memory and place, thirdly, take a particular relevance in relation to *(post)colonialism*. Said (2000) reviews that by emphasizing the hold of both memory and place on the desire for mapping, conquest and annexation of territory, but also the more subtle and complex unending struggle over territory, which necessarily involves overlapping memories, narrations and material environments.

It should be obvious from the above, that time in this connection is neither homogeneous nor a continuous historical narrative. It is rather a multitude of events and qualitative different moments (de Certeau 1984). But in accordance with the conception of place at stake here, place should neither be seen as a passive container for preserving or storing memories:

> Places mix times into the present, mixing orders of virtual and actual. Equally, with the different times coexisting, 'places' are not unitary spaces and times but include subterranean landscapes of fragmented spaces. Rather than being an immensity of past from which we select, we have moments irrupting through places to bring the past into contact with the present. (Crang and Travlou 2001)

What we are dealing with in relation to memory is a dynamic sense of space-time in which different times are imbricated in places. Places do not offer unification or stability but include different times in their discontinuous processes of becoming.

Conclusion: Place as Encounters

The account of place I have tried to come to terms with in this chapter approaches it through space-time configurations or, maybe more illustrative, what Lefebvre (1992) calls *rhythms*. Rhythms, in this sense, are spatio-temporal flows of living bodies and their internal and external relationships. Different (and multiscalar) practices are characterized by different rhythms, between which there are dynamic relations and interferences. Place, from that point of view, becomes a locus of encounters, the outcome of multiple becomings.

> …'here' is no more (and no less) than our encounter, and what is made of it. It is, irretrievably, here *and* now. It won't be the same 'here' when it is no longer now. (Massey 2005, 139)

Places are meeting points, moments or conjunctures, where social practices and trajectories, spatial narratives and moving or fixed materialities meet up and form configurations that are continuously under transformation and negotiation. Such space-time configurations are never bounded, but the succession of meetings and returns, the accumulation of relations and encounters, build up a history that lends continuity and identity to place. Places are encounters marked by openness and change but not without material, social and cultural duration.

Like other present-day ones, this account of place is *relational*, in my case giving priority to the social content of the relations through which place is constructed. Different suggestive metaphors have been introduced to take in this relationality. Thrift (1999) suggests *passings* to advance an understanding of places 'as taking shape only in their passing'. I find that too elusive and missing what Harvey (1996)

– maybe too much to the opposite site – calls *permanences*, hereby describing places as entities achieving stability for a time, through some bounding and internal ordering of processes creating space. The more helpful suggestion might be Massey's (2005) (rather Heideggerian) notion of *throwntogetherness*, emphasizing the unavoidable challenge of the negotiation of multiplicity, the sheer fact of having to get on together.

This notion draws attention to the fact that place as encounters is always also encountering 'the Other', that places cannot be 'purified'. Attempts to do so are exactly a consequence of this fact. At one level, we can see such encounters as face-to-face meetings, but encounters are also always mediated. They presuppose other faces, other encounters of facing, other bodies, other spaces, and other times. Existing spatialities and temporalities – and embodiments, emotions, narratives and memories – are translated into every encounter as formative layers of hybridity. This inevitably also translates into politics of place, involving negotiations of openness and closure and the possibility of what Amin (2004) calls a heterotopic sense of place.

References

Ahmed, S. (2000), *Strange Encounters. Embodied Others in Post-coloniality* (London and New York: Routledge).

Amin, A. (2004), 'Regions Unbound: Towards a New Politics of Place', *Geografiska Annaler* 86B:1, 33–45.

Augé, M. (1995), *Non-Places: Introduction to an Anthropology of Supermodernity* (London and New York: Verso).

Bærenholdt, J.O. and Simonsen, K. (eds) (2004), *Space Odysseys. Spatiality and Social Relations in the 21st Century* (Aldershot: Ashgate).

Bourdieu, P. (1977), *Outline of a Theory of Practice* (Cambridge: Cambridge University Press).

—— (1990), *The Logic of Practice* (Cambridge: Polity Press).

Casey, E.S. (1993), *Getting back into Place* (Bloomimgton/Indianopolis: Indiana University Press).

de Certeau, M. (1984), *The Practice of Everyday Life* (Berkeley: University of California Press).

Crang, M. and Travlou, P.S. (2001), 'The City and the Topologies of Memory', *Environment and Planning D: Society and Space* 19, 161–77.

Cresswell, T. (2004), *Place – A Short Introduction* (Oxford: Blackwell).

Crossley, N. (1996), *Intersubjectivity. The Fabric of Social Becoming* (London: Sage).

Dreyfus, H. and Hall, H. (eds) (1992), *Heidegger. A Critical Reader* (Oxford: Blackwell).

Entrikin, J.N. (1984), 'Carl O. Sauer, Philosopher in Spite of Himself', *Geographical Review* 74, 387–408.

—— (1991), *The Betweenness of Place: Towards a Geography of Modernity* (London: Macmillan).

Guignon, C. (2003), 'Moods in Heidegger's *Being and Time*', in Solomon (ed.).

Halbwachs, M. (1992), *On Collective Memory* (Chicago and London: Chicago University Press).

Harvey, D. (1989), *The Condition of Postmodernity* (Oxford: Blackwell).

—— (1996), *Justice, Nature and the Geography of Difference* (Oxford: Blackwell).

Heidegger, M. (1962), *Being and Time* (Oxford: Blackwell).

Hobsbawm, E. and Ranger, T. (eds) (1983), *The Invention of Tradition* (Cambridge: Cambridge University Press).

Hoelscher, S. and Alderman, D.H. (2004), 'Memory and Place: Geographies of Critical Relationship', *Social and Cultural Geography*, 5:3, 347–57.

Jameson, F. (1991), *Postmodernism, or, the Cultural Logic of Late Capitalism* (Durham: Duke University Press).

Laclau, E. (1990), *New Reflections on the Revolution of our Time* (London: Verso).

Lefebvre, H. (1991), *The Production of Space* (Oxford: Blackwell).

—— (1992), *Élements de rythmanalyse: Introduction à la connaissance de rythmes* (Paris: Syllepse).

Lefebvre, H. (1996) (edited by Kofman, E. and Lebas, E.), *Writings on Cities* (Oxford: Blackwell).

Massey, D. (1994), *Space, Place and Gender* (Oxford: Polity Press).

—— (2005), *For Space* (London: Sage).

Massey, D. and Allen, J. (eds) (1984), *Geography Matters! A Reader* (Cambridge: Cambridge University Press).

Massey, D., Allen, J. and Sarre, P. (eds) (1999), *Human Geography Today* (Cambridge: Polity Press).

Merleau-Ponty, M. (1962), *Phenomenology of Perception* (London: Routledge and Kegan Paul).

—— (1968), *The Visible and the Invisible* (Evanstone, IL: Northwestern University Press).

Öhman, J. and Simonsen, S. (eds) (2003), *Voices from the North. New Trends in Nordic Human Geography* (Aldershot: Ashgate).

Paasi, A. (1986), 'The Institutionalisation of Regions: A Theoretical Framework for Understanding the Emergence of Regions and the Constitution of Regional Identity', *Fennia* 164, 105–46.

—— (2002), 'Place and Region: Regional Worlds and Words', *Progress in Human Geography* 26:6, 802–11.

Pred, A. (1984), 'Place as Historically Contingent Process: Structuration and the Time – Geography of becoming Places', *Annals of the Association of American Geographers* 74:2, 279–97.

Relph, E. (1976), *Place and Placelessness* (London: Pion).

Ricoeur, P. (1984), *Time and Narrative* vol 1 (Chicago: University of Chicago Press).

Rose, G. (1993), *Feminism and Geography: The Limits of Geographical Knowledge* (Cambridge: Polity Press).

Said, E.W. (2000), 'Invention, Memory, and Place', *Critical Inquiry* 26, 175–93.

Schatzki, T. (2002), *The Site of the Social: A Philosophical Account of the Constitution of Social Life and Change* (University Park, Pennsylvania: The Pennsylvania State University Press).

Simonsen, K. (1995), 'Sted, køn og social praksis – om en "kønnet" forståelse af stedsbegrebet', *Nordisk Samhällsgeografisk Tidskrift* 21, 22–32.

—— (2003), 'The Embodied City – from Bodily Practice to Urban Life', in Öhman and Simonsen (eds).

—— (2004), 'Spatiality, Temporality and the Construction of the City', in Bærenholdt and Simonsen (eds).

Solomon, R.C. (ed.) (2003), *What is an Emotion? Classic and Contemporary Readings* (New York/Oxford: Oxford University Press).

Thrift, N. (1996), *Spatial Formations* (London, Thousand Oaks and New Delhi: Sage).

—— (1999), 'Steps to an Ecology of Place', in Massey, Allen and Sarre (eds).

Thrift, N. and Dewsbury, J-D. (2000), 'Dead Geographies – and how to make them live', *Environment and Planning D: Society and Space* 18, 411–32.

Chapter 3

Being Along: Place, Time and Movement among Sámi People

Nuccio Mazzullo and Tim Ingold

Places Along the Way

During fieldwork in Finnish Lapland, one of us – Nuccio Mazzullo[1] – listened in on a conversation between a pair of friends, both of them Sámi, prompted by his questions about how difficult it was in the past to travel during the summer. They answered by explaining about the importance of rivers and lakes. One of the pair, who was from the village of Kuttura, recounted the journey he used to make to another village upriver, called Matti of Ivalo (Ivalon Matti).[2] As he did so, he ran through a series of place-names along the roughly ten-kilometre stretch of the Ivalo River from Kuttura to Matti. The names, told in Finnish, generally comprised two parts. The first part of each name refers either to an occurrence (such as an encounter with a wild reindeer, an elk or a fox, or a fishing episode) or to some characteristic of the shore (being rocky, bending, covered in lichen), while the second part describes either the strength of the river current or the joining of tributaries. Thus, the suvanto or 'quiet waters' is a stretch on a river where the water flows relatively slowly, and the niva or 'small rapids' is a stretch where the stream is relatively strong. The koski are normal rapids, the köngäs are big, agitated and frothy rapids, the kärrää (from the Sámi garas or garra, meaning harsh or hard) are even stronger rapids, and the suu is the river mouth. Oja is a small stream or brook, but can also mean a ditch cut for drainage.

---

1   The Sámi are indigenous inhabitants of the northernmost regions of Norway, Sweden, Finland and Russia's Kola Peninsula, and have traditionally drawn a livelihood from hunting, gathering and fishing and – from around the sixteenth century – reindeer herding and small-farming. In Finland, the Sámi currently number about 6,000 people. The fieldwork on which this chapter is based was carried out by Nuccio Mazzullo in the Inari region of Finnish Lapland between 1995 and 1997, and intermittently since then. Much of the ethnographic material is drawn from Chapters 3, 4 and 5 of the resulting doctoral dissertation (Mazzullo 2005), which was written under the supervision of Tim Ingold. Ingold's contribution to this chapter lies primarily in the further development of some of the theoretical ideas. These ideas are set out more fully in Ingold (2007, 72–103). Mazzullo acknowledges the award of a three-year studentship from the UK Economic and Social Research Council (ESRC) in 1995–98. Ingold acknowledges the award of an ESRC Professorial Fellowship (2005–8), which has afforded him the time to work on this chapter.

2   Here, and in what follows, words in the Finnish language are underlined, while words in Sámi are written in italics, except when listed.

Here is the list of names, from Kuttura to Matti of Ivalo, in the order in which they were recounted:

Peurasuvanto (wild reindeer – quiet waters)
Lippasuvanto (trolling bait – quiet waters)
Hirviniva (elk – small rapids)
Hirvisuvanto (elk – quiet waters)
Harriniva (grayling – small rapids)
Mukkakoski (bend – rapids)
Mukkasuvanto (bend – quiet waters)
Suljuniva (slimy – small rapids)
Saarikosken suvanto alla (island rapids – quiet waters below)
Saarikoski (island – rapids)
Jäkäläkoski (lichen – rapids)
Kivikoski (rocky – rapids)
Kivisuvanto (rocky – quiet waters)
Järnänköngäs (Järnä river – strong rapids)
Järnänsuu (Järnä river mouth)
Karvaniva (hairy – small rapids)
Kuplettioja (comic song – brook)
Repojoensuu kärrää (Fox river mouth – very strong rapids)
Repojoen suvanto (Fox river – quiet waters)
Kaunisniva (beautiful – small rapids)
Sormusnivat (ring – small rapids)
Sormussuvanto (ring – quiet waters)
Ivalon Matti nivat (Matti of Ivalo – small rapids)
Ivalon Matti suvanto (Matti of Ivalo – quiet waters)

To list these names in sequence is, in effect, to render an account of the journey, and at the same time to bind the flow of water through the land with the paths of humans and animals and the activities that take place along them. It is evident that the narrator is 'travelling', in his list, against the stream: thus, setting off from his home in Kuttura, which is situated near the first place on the list, he encounters the quiet waters lying below island rapids before he comes to the rapids themselves. Another, similar example of sequencing comes from the following interview with a Sámi person who grew up in the village of Lisma. What follows is an extract from a transcription of the interview, translated from the original Finnish:

> When we left from Lisma on our way to Nunnanen, the winter path went through all the large swamps. But in summer, when we travelled on foot, nobody wanted to walk through those boggy swamps, so we would go over the dry land. I can well remember almost all the swamps from Lisma, but not really as far as Nunnanen because I am not completely familiar with the Nunnanen area. But nearer Nunnanen every swamp has a name from when you leave Lisma…

Mazzullo interrupts to ask: Are they on the map?

Well, not all the swamp names are marked on the map…there is Karpalojänkä [cranberry swamp], Pulttijänkä [bolt swamp], Ahkiojänkä [Sámi sledge swamp]. Then from Ahkiojänkä starts Kuolpanoja [dry peaty forest brook], along which the winter path went. Then on one side of Kultakuru [gold gully] there begins a large swamp that leads almost to Nunnanen, but what was the name…oh there it is, it was Skierrivuoma [*skier'ri*, dwarf birch], at least that was one of them, but this Skierrivuoma [dwarf birch marsh] is on the northern side…what were the names of the other swamps, I don't remember. It's there close to Nunnanen, and then there, a couple of kilometres from Lisma, there was another swamp but I don't remember its name…it was Korva…it was Kovakorva [hard ear swamp]. I don't know whether it had any more particular name, that first swamp. But all those bigger swamps had names, and if you think of the distance from Lisma and Repojoki, which is 24 kilometres, well there were a lot of swamps you had to cross. One of them is Kahden Autsin Välinen Jänkä [the swamp between the two Autsi hills]. Then there is Heinäjänkä [hay-swamp]. Then there is Spiinijänkkä Muukka. Then there is the Kahden Nirrusala Välinen Jänkä [the swamp between the two Nirrusala hills], then there should be…it is actually Spiinijänkä [pig swamp, from the Sámi *spiidni*, 'pig']. It could be that Hattivuoma is the last one…and then there is Ponkun Räme…[the marsh of Ponku]. (Transcribed and translated from original recording of interview, 21 August 1997)

In both these and many other examples, narrators would re-enact their journeys as they recollected the names of the places through which they used to pass. As they did so, even while sitting indoors, they would cast their eyes towards various corners of the room. Their heads would turn and their hands would point in different directions. In their bodily gestures they rehearsed the same movements that they would have made during the journey. Thus the act of naming was in itself part of this narrative performance. Every name would be recalled not as an isolated sign affixed to a specific location, as it might appear on a cartographic map, but as one episode or topic of a story – that of the journey recounted. Thus, recalling the names of places is tantamount to the act of travelling through them, in the imagination if not on the ground. Sámi people would readily recall names as they travelled on the land, or as they described the journeys they had made, but if asked for the name of such and such a place, as indicated by a particular location on the map, they would never give a straight answer but rather find some means of evading the question. For these names mean nothing on their own. In short, just as places can only occur along paths of movement, so their names can only be told in sequence, in the narrative re-enactment of journeys actually made.

Remembering these sequences serves an important practical purpose. When travelling on the land, it is crucial to be able to find one's way back. For this purpose the traveller would pay particular attention to such features as swamps, clearings, hills, different types of vegetation and clusters of trees encountered along the way. Also important for orientation are standing dead trees (*soarvi*), big or peculiarly shaped or positioned rocks, and so on. Of course the salience of these features depends upon the season of the year. In summer it is important to remember the direction of the stream and the type of watercourse. One has to know how the major watercourses are connected to the various lakes, and similarly also the connections between swamps, highlands and hill chains. The height of the sun over the horizon can be used for orientation during the night in the summer months, when the stars are not visible, but this is done in conjunction with the observation of other features of the landscape, such as the direction or concentration of tree branches and anthills. Though Sámi people may nowadays carry maps with them, or even sophisticated GPS devices, these are seldom if ever used for the purposes of orientation. 'I never use maps when I go to the forest', Sámi friends would say. 'Maps are for tourists!'

At Home in the Forest

Now philosophers assure us, time and again, that as earthbound creatures we can only live, and know, *in* places (Casey 1996, 18). An example of this way of thinking may be found in the work of J. E. Malpas. For him, places exist on many levels in a nested series, so that whatever level we may select, a place is liable both to contain a number of lower level places and to be contained, alongside other places at that level, within a higher level place. Thus 'places always open up to disclose other places within them…while from within any particular place one can always look outwards to find oneself within some much larger expanse' (1999, 170–71). To leave one's place of abode, then, is to take a step upwards, from a smaller, more exclusive place to a larger, more inclusive one, and thence to a still larger and yet more inclusive one. The higher one climbs, however, the further removed one feels from the groundedness of a place, and the more drawn to an abstract sense of space. Conversely, the return trip homeward takes the traveller on a downward movement, through these levels, from space back to place (Malpas 1999, 171). It is as though life can be lived authentically only at the base of a vortex, from which there is no escape but to lift off the ground of bodily experience, upwards and outwards, towards ever higher levels of abstraction.

It was not, however, from such a base of dwelling that Sámi people looked up and out upon the wider world – or not, at least, until relatively recent times, when they took up residence in permanent houses. Formerly – that is until the 1930s, well within living memory of the older generation – they had lived in conical tent dwellings. The principal components of the dwelling (*lávvu*) are the supporting poles, which were either left standing (particularly in the tundra, where trees are scarce) or cut on the spot from small trees, and the two canvases (*loavdagat*) that covered the structure and always travelled with its inhabitants. Once the wooden frame was wrapped with the canvases and the fire at the centre of the tent had been lit, the *lávvu* would come alive, busied by all the domestic tasks associated with it. Even today, the *lávvu* is used by reindeer herders when they are working out in the terrain. As soon as the reindeer men had put up the *lávvu*, a place (*báiki*) would be formed, and a fire would be lit and coffee quickly brewed in practical affirmation of the fact. From the term *lávvu* is also derived the verb *lávvudit*, which is used to describe where one has been and what one has done. It can be translated as 'to camp', but is used only for camping in a familiar place.

Yet while the *lávvu*, brought to life by its inhabitants, gathers their activities into a focus of dwelling, the forest in which it stands is no less homely. Thus a Sámi person does not leave home to go to the forest. The image famously invoked by Martin Heidegger (1971, 154) of the space of dwelling as a *clearing*, though appropriate enough for Finnish settlers in the North who brought with them a way of life based on farming and forestry, fails to capture the Sámi sense of what it means to be at home. For the Sámi, home is open-ended and unbounded. Far from being a narrowly circumscribed clearing in the woods – a low-level place encompassed by the higher-level, more expansive place of the forest – home comprises the entire web of trails that converge upon the dwelling. For what is called the forest, *meahcci* (equivalent to the Finnish <u>metsä</u>), is understood not so much as a tree-covered expanse as a texture densely interwoven from the paths along which people carry on their activities of herding, fishing, berry-gathering and so on. Indeed the forest may, as in the tundra, be almost or entirely treeless. If you were to ask where a person is, the answer would come back in terms of what they were doing: thus they may be in the 'herding forest', the 'fishing forest' or the 'berrying forest', referring in each case to the paths that people would take in that particular activity (see Figure 3.1).

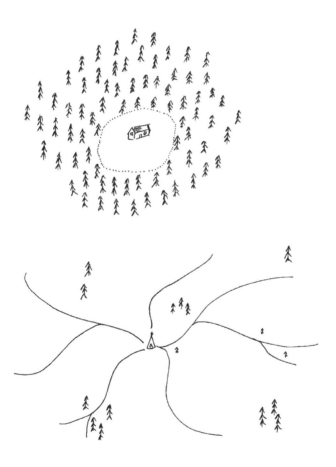

Figure 3.1   The experienced space of dwelling

Source: drawing by Tim Ingold.

The space of dwelling, in the experience of Finnish settlers (above) and indigenous Sámi (below). For settlers, the dwelling house stands in a clearing in the forest (metsä), and the edge of the woods marks the limits of home. For Sámi people, by contrast, the tent-dwelling or lávvu gathers together the trails that people follow in their everyday dwelling activities. The forest (meahcci), which may be tree-covered only here and there, is the sum of these trails, interwoven with trails of all its other inhabitants, human and non-human. Home and forest are one.

Following such paths the individual remains as grounded, and the experience remains as embodied, within the dwelling. And what each trail progressively discloses is not any wider expanse but a series of vistas, occlusions and transitions as, along the way, horizons open up ahead and are in turn closed off behind. As we have seen, every path is remembered as a sequence of such openings, as in the case of the swamps on the way from the village of Lisma, running between hills on either side. The river is likewise remembered as a sequence of 'waters' such as those between the villages of Kuttura and Matti of Ivalo. Making his or her way, by foot or rowboat, the traveller is always somewhere, but every somewhere is on the way to somewhere else, every place on a path. In the past, the Sámi used to set up storage places along their trails so that they could retrieve the items they needed in accordance with the characteristics of the place and the season, without having to carry everything with them. Some have described the transition to living in permanent houses as equivalent to being locked up in one room of a castle containing a hundred rooms. But if the Sámi's home was his castle, that castle was not really an ensemble of enclosed rooms but a labyrinth of passageways.

Our contention, then, is that life – at least for Sámi people – is lived not *in* places but *along* paths. Here we take issue with philosophers such as Edward Casey. Place, according to Casey, is 'at once the limit and the condition of all that exists…To be is to be in place' (Casey 1993, 15–16). We are in place, he asserts, because we exist as *embodied* beings. The bodily experience of the Sámi person is not, however, confined within a somatic shell but rather extends as it grows and is laid down along the multiple paths of an individual's movement through the forest. Thus to be, for a Sámi, is to be not *in* but *along*. The path, not the place, is the primary condition of being, or rather of becoming. Places are formed through movement, through the endless current of coming and going from and to places elsewhere. Such a coming-and-going movement is place-*binding*, but it is not place-*bound*.[3] And to go from place to place means going along – always grounded in an immediate perceptual engagement with one's surroundings – not up and out to higher, less engaged levels and back down again. Places, then, do not so much exist as *occur* – they are stations along the byways of life. Instead of saying that people exist in places, it would be closer to the experience of the Sámi to say that places occur along the life-paths of persons.

---

3   Note that bindings are not boundaries. Threads that are bound together *in* a knot are not bounded *by* it. On the contrary: they trail beyond it, only to become caught up with other threads in other knots. It is the same with paths and places (Ingold 2007, 100).

The Conjunctures of Life

Now movement takes time, and if to *be* is to be *along* a path, then being must be intrinsically temporal. Writing in the late 1940s, the celebrated Finnish ethnologist T. I. Itkonen observed that 'the passage of time is perceived through the alternation of the day and the night, during the day through the sun, and during the night especially through the Great Dipper, the Pleiades, and the morning star' (Itkonen 1948, 490, our translation from Finnish). To signify a minute or a few moments, Sámi people would use an expression like (for example) *pii'po-tsâhk'ke'tâm-pod'dâ*, 'the lighting of a pipe'. If the period was longer, then they would use the expression *käf'fe-vuoš'šâm-pod'dâ*, the time needed for cooking coffee, though Itkonen noted that among reindeer herders this could mean some two or three hours (Itkonen 1948, 490–1). During his fieldwork, Mazzullo also found that this latter expression was highly dependent on context. If somebody used it in a village it meant about fifteen minutes or less, whereas if it was used in the forest or during the reindeer separation or earmarking then it meant a period of a couple of hours.

While travelling, if anyone were to ask at what time they would be arriving or returning, only the vaguest indications would ever be given. The length of a journey would be expressed as the time needed to perform all the tasks necessary for people to reach a given destination. This time is variable. Thus it might be known that going through one particular place is bound to be tiresome, while other places might be particularly easy; since the degree of difficulty or easiness could only be assessed on the spot, however, any attempt at greater precision could only be a matter of guesswork. In normal life, and in activities unrelated to official public time-tabling, people are relaxed and flexible. Rather than 'being late' in the completion of particular tasks, or for an appointment, one would draw attention to some task having been 'particularly difficult', or simply say that one had 'something more compelling to do' or just 'was unable to attend'. No-one expects apologies or justifications for not turning up at a given time or day, though some teasing may occur. However, people are always ready to do something else instead.

This attitude towards punctuality is perfectly in tune with the personal autonomy that is so highly valued among the Sámi. If you are worried about doing something on time, people will be quick to reassure you that 'if you can't do it this spring, then do it next spring'. Although such remarks could be interpreted as attesting to a notion of cyclical time, we are more inclined to see them as an injunction to ignore what Pierre Bourdieu (1963, 58) calls the 'tyranny of the clock'. In itself, the clock cannot act like a tyrant. It is nothing more than a mechanical device. It only becomes an instrument of tyranny when it is hooked up to structures of political control and subordination. If it is not connected to such structures then the clock exerts no tyranny at all. When Sámi people go to the forest, or during the reindeer separation and the earmarking of calves, no-one would ever look at their watches. In these northerly latitudes there is not much difference in luminosity between day and night in either winter or summer. In winter, the reindeer separation would be started in pitch darkness and it would still be dark when the work was finished. Likewise in summer, the earmarking would commence late at night, though in full daylight, and finish some time in the morning. Only then did everyone go to the huts to sleep.

No-one ever acknowledged having worked for twelve or fifteen hours, nor was any distinction made between a time to sleep and a time to work. During breaks, people used to take a nap anytime they felt they needed one.

However, unlike the Kabyle peasants, whose unhurried, day-to-day routine has been so eloquently described by Bourdieu, herding work has more of a 'stop-go' character, with periods of very intense, hasty activity interspersed with periods of 'alert idleness'. If there is anything cyclical in the Sámi perception of time then it lies in the periodic 'coming together' and 'moving apart' of circumstances whose conjunction is auspicious for carrying out particular tasks. It is not time that is crucial so much, as the *timing*. The Sámi view, clearly, is that timing has to be judged not by the clock but by a particular conjunction of auspicious circumstances. Harvey Feit reports a similar view among the Waswanipi Cree, indigenous hunters in Northeastern Canada. The Cree, according to Feit, believe that 'humans do not ultimately control life, but intimately and respectfully link their thought and action to those powerful beings who create the *conjunctures of life*' (Feit 1994, 435, our emphasis). Likewise, for the Sámi, it is at those moments when everything comes together, in the conjunctures of life, that the time is right to make a move. At such moments, which may not be predictable, one has to be ready to move *at once* and quickly, otherwise the conjunction may pass (as everything else is on the move as well – the herd, the weather, other people) and one may lose the opportunity.

Sámi people have a flexible attitude to the relation between the time needed to accomplish a task and individual styles of accomplishment. They acknowledge that people differ in both the pace and the ways in which they accomplish similar tasks. Unless absolutely necessary people would not expect, nor would they be expected, to set out a schedule within which things should be done. The only master of deadlines is the weather. If a trip to the pasture had to be made before weather conditions deteriorated to such an extent that it would become impossible, or if a lull in bad weather conditions allowed it to go ahead, then it would start without delay. As these examples show, what is important for Sámi people is not that they should be somewhere at a particular time but that they should be ready to be there at *any* time – which could be in an hour or two, or in several days or weeks – whenever the circumstances are such as to require it. Even when resting, therefore, one has to be constantly alert. In short, what counts is not punctuality but *readiness*, not the precise targeting of a point in time but a continual monitoring of the way things are going, in a world in which everyone and everything is in movement, each at their own pace, along alternately converging and diverging paths.

Beyond Landscape and Taskscape

In Bourdieu's study of the Kabyle peasants of Algeria, which formed the ethnographic foundation for the subsequent development of his theory of practice (Bourdieu 1990), the perception of time is closely associated with practical activity:

> duration and space are described by reference to the performance of a concrete task; the unit of duration is the time one needs to do a job, to work a piece of land with a pair of oxen. Equally, space is evaluated in terms of duration, or better, by reference

to the activity occupying a definite lapse of time, for example, a day at the plough or a day's walk. (Bourdieu 1963, 60)

This passage is closely in tune with an approach to the temporality of the landscape that one of us – Ingold – has developed in earlier work. His argument was that the rhythmic structure of social time emerges not only from the interweaving and mutual responsiveness of human movements: it also resonates to the cycles of the non-human environment. For traditionally people had to fall in with the rhythms of their environment: with the winds, the tides, the needs of domestic animals, the alternations of day and night, the seasons and so on, in accordance with what the environment afforded for conducting their daily tasks (Ingold 1995, 9).

In appealing to the notion of *task*, both Bourdieu and Ingold highlight the aspect of people's practical involvement with their lived worlds. Ingold defines the task as 'any practical operation, carried out by a skilled agent in an environment, as part of his or her normal business of life' (Ingold 1993, 158). But as he goes on to show, tasks are not performed in isolation, for 'every task takes its meaning from its position within an ensemble of tasks, performed in series or in parallel, and usually by many people working together' (Ingold 1993, 158). Ingold calls this ensemble the *taskscape*. Hence, just as every place is situated within a landscape, every task is situated within a taskscape. If tasks are carried out by skilled agents, however, not every agent is necessarily a human person. Along with other indigenous peoples of the northern circumpolar region, Sámi people tend to treat the various constituents of their environment in ways that draw no absolute distinction between human and non-human realms (Ingold 2000, 47). Nor do they draw a clear line between animate beings and what we might regard as inanimate things. For animacy is not regarded as an intrinsic property of things, but rather as immanent in the entire texture of relations within which all the constituents of the environment – from rocks to trees to animals and humans – are woven together.

Now if, as we have argued, every constituent is instantiated in the world not *in* place but *along* a path – as a line of movement – and if every place occurs along a path as a moment in its unfolding, then it is no longer possible to separate out the landscape from the taskscape. From an indigenous perspective, the landscape is not an inanimate backdrop or substrate, against or on which animate beings perform their respective tasks. People and animals are as much part of the landscape as are rocks and trees, and rocks and trees are responsive to the activities of animals and people just as the latter respond in their tasks of life to the activities of one another. Mutual responsiveness is based on interaction, and interaction implies movement. For, fundamentally, everything is suspended in movement. In this movement, people and things are continually coming together, merging and splitting apart again, only to come together once more in different combinations. In the world of human beings, this is the principle of bilateral kinship. But it operates equally in the non-human domain, and in the domain of relationships between humans and non-humans. The importance of movement is continually emphasized among the Sámi, and Mazzullo was constantly made aware of it during his fieldwork.

We have seen how, for Sámi people, movement is the key to life. We have also seen, on a more theoretical level, how movement is intimately related to the performance of tasks and how such performance, in turn, is foundational to the perception of time. Our contention is that movement is the underlying principle that animates the Sámi world. This principle of movement should not be confused with the physics of locomotion. By locomotion we mean the mechanical propulsion of an object of some kind from point to point across a surface that is already laid out – as for example of a piece across a gaming board. To be able to move, in this sense, the object must be disconnected from the surface. Conversely, to be connected to the surface it must be fixed in place. When people who live by hunting and gathering or by herding have been called 'nomadic' by outsiders, their movement has generally been understood in this mechanical sense, with the corollary that only by settling down – by becoming 'fixed' to the land surface – can they establish any real connection with it. Nomadism, in this sense, necessarily implies a connection to the land that is tenuous or non-existent.

From the perspective of the people themselves, however, it is precisely through their movements that they are entwined with the land. These are movements not of being but of becoming – of growth and self-renewal along ways of life (Ingold 2000, 146). As we have already shown, the bodily experience of Sámi people is laid down as they thread their ways along paths *through* the land, contributing through their movements to its ever-evolving weave. The land itself – or in Sámi terms the forest – is not a ready-made surface but a tangle of pathways woven from the movements of all its manifold inhabitants, human and non-human. The corollary of this, of course, is that the effect of enforced sedentarization is to rupture people's bonds with the land, such that these become more rather than less tenuous. Constrained to reside in one spot, they can no longer inscribe their lives into the land as they used to do. Nevertheless, as they 'go about' in the terrain in their everyday tasks, responsive to the movements of the herds and alert to the conjunctures of wind and weather, the forest is still 'home' to them in a way that it is not for those whose lives are lived within the confines of a clearing.

Ahead of his time, Gregory Bateson pioneered the attempt to draw a full, encompassing approach to human environmental relationships in his *Steps to an Ecology of Mind* (1973). In this collection of essays he bases his theory of perception on the notion of the cybernetic circulation of information. Rejecting theories that set the skin as the outer boundary of the mind (1973, 428–9), Bateson saw 'mind' as immanent in the total system of pathways along which information flows from the person into the environment and back again. Somewhat analogously, we have shown how 'home' for the Sámi is not confined within the canvas skins (*loavdagat*) of the tent dwelling but is immanent in the meshwork of trails that take people around and about in the environment of the forest. Now the generative source of information, for Bateson, lies in 'difference' and all difference implies change of some kind. Not all change, however, implies difference. Mere motion, 'the simplest and most familiar form of change' (1973, 254), generates no information. It only does so when it 'makes a difference' in terms of its consequences. In a footnote to one of his essays Bateson calls the type of movement that makes a difference, in this sense, the 'movement within the movement' – or in a word, 'metamovement' (1973, 252). An object propelled from point to point across a surface may be said to have moved, but as long as this displacement has not altered its nature, there has been no movement in its motion. A living being, however, is one whose movement is itself a process of becoming. That becoming is its metamovement. And it is because of this metamovement that no living being is quite the same, on arrival at a place, as when setting out.

Drawing on Bateson's idea, we refer to this generative principle of life as *metakinesis*. This notion of metakinesis transcends our conventional dichotomies between statics and dynamics, between synchrony and diachrony, and between space and time. We have already shown that, for the Sámi and other indigenous peoples, no clear-cut line can be drawn between landscape and taskscape. It is not possible to separate the interrelated activities of skilled agents that constitute their taskscape from the world – the landscape – in which they operate. In order to overcome the opposition between landscape and taskscape we therefore propose that the world inhabited by the Sámi is a *metakinesis-scape*.[4] In essence, the metakinesis-scape is constituted of the convergence of all the different purposeful movements by which agents of diverse kinds interact in different terms and with different modalities, ultimately forming the very life-world that enfolds them. The Sámi life-world is a metakinesis-scape, since it comprises an entire field of rhythmic interrelations (a *scape*) between movements of terrestrial travel and bodily gesture (*kinesis*) that are themselves (*meta*) movements of becoming that make a difference to both mobile beings and the world they inhabit.

References

Bateson, G. (1973), *Steps to an Ecology of Mind* (London: Paladin).

Bourdieu, P. (1963), 'The Attitude of the Algerian Peasant toward Time', in Pitt-Rivers (ed.) (q.v.).

—— (1990), *The Logic of Practice* (Cambridge: Polity Press).

Burch, E.S. Jr. and Ellanna, L. (eds) (1996), *Key Issues in Hunter-Gatherer Research* (Oxford: Berg).

Casey, E.S. (1993), *Getting Back Into Place* (Bloomington: Indiana University Press).

—— (1996), 'How to Get from Space to Place in a Fairly Short Stretch of Time: Phenomenological Prolegomena', in Feld and Basso (eds) (q.v.).

Feit, H. (1994), 'The Enduring Pursuit: Land, Time and Social Relationship in Anthropological Models of the Hunter-Gatherers and in Sub-Arctic Hunters' Images', in Burch and Ellanna (eds).

Feld, S. and Basso, K.H. (eds) (1996), *Senses of Place* (Santa Fe, NM: School of American Research Press).

Heidegger, M. (1971), *Poetry, Language, Thought*, trans. A. Hofstadter (New York: Harper and Row).

Ingold, T. (1993), 'The Temporality of the Landscape', *World Archaeology* 25:2, 152–74.

---

4   The term is admittedly cumbersome, and we offer it more as a summary of our argument than as a suggestion for further use.

—— (1995), 'Work, Time and Industry', *Time and Society* 4:1, 5–28.

—— (2000), *The Perception of the Environment: Essays on Livelihood, Dwelling and Skill* (London: Routledge).

—— (2007), *Lines: A Brief History* (London: Routledge).

Itkonen, T.I. (1948), *Suomen Lappalaiset Vuoteen 1945* [The Finnish Lapps up to the year 1945], 2 Vols. (Porvoo: WSOY).

Malpas, J.E. (1999), *Place and Experience: A Philosophical Topography* (Cambridge: Cambridge University Press).

Mazzullo, N. (2005), *Perception, Memory and Environment among Sámi People in Northeastern Finland* (Unpublished doctoral dissertation, University of Manchester).

Pitt-Rivers, J. (ed.), *Mediterranean Countrymen: Essays in the Social Anthropology of the Mediterranean* (Paris: Mouton).

Chapter 4

Sensing Places:
The Ethics of Authentic Place

Inger Birkeland

Introduction

When talking about tourists' travel to other places we often focus on their experiences of places in the external world, in terms of natural attractions, heritage sites or spectacles in one sense or another. Places can, however, also be seen as existing in an internal world, and travel can be viewed as place-making in both external and internal worlds. In this chapter I will focus on the nature of place more generally, on the individual and authentic experience of place more specifically, and finally ways to place-making that are more sustainable than those found in modernity's production of placeless places.

I would like to discuss this by first telling the story of a real journey based on ethnographic research among travellers to the North Cape (Birkeland 2005). Ten years ago, a Spanish woman, 'Sofia', walked the distance from Oslo to the North Cape alone, by foot. Sofia started walking in late winter in 1997 and arrived at the North Cape in late summer the same year. Her journey differs from most tourist travels made to the Nordic north in many obvious ways. Let us focus here on her particular method for travelling, which was walking. For me it has been interesting to look at her walking as a method for transcending the dualist split between self and place and arrive at a way of seeing selves and places as reciprocally related. What I will do in the following, is to use Sofia's experiences of walking to sketch a sense of place that takes the form of a reflexive journey from a disembodied, abstract and geometrical sense of place to an embodied and animated sense of place. I will show how a return to the senses also represents a departure towards authentic place-making and environmental humility.

From Oslo to North Cape by Foot

Sofia lived by herself in a city in Spain and had a good job in a computer company. In the mid-1990s Sofia had come to a point in her life when she felt that what she did was not meaningful. She did not know what to do, but she knew she could not continue as she did. She decided that she had to quit her job. Her friends and her family felt that she had made a huge mistake, and deeply worried her mother confronted her and asked: Have you lost your North? Sofia then explained to me that this is a typical saying in Spain which means something like: Have you lost your mind? The North is associated with a secure and stable point of reference, in a person's map. If one knows one's north, one knows what is most important, one knows where one is, and thus who one is. If you don't know where your north is, you do not know who you are. If one knows where the north is, one also knows where one's east, south and west is. We can say that the north refers to a point in a person's life-world in a phenomenological sense which is a subjective geography, the landscape of subjectivity. We can thus imagine subjectivity as a place-world.

I guess many people can recognize from their own life experiences feelings of lacking life direction or loss of reference or meaning, but very few people would go so far as to take the meaning of the north point as literal as Sofia did. Sofia told me that she decided to go to the north in order to find her north point, to find her personal north point. It was something that compelled her to travel northwards, and she knew she had to do this journey all by herself. She moved to Norway, where she found work and stayed for some months, but feeling very depressed, she knew she had not found her north yet. Then, one morning in early 1997 she suddenly knew what to do. A new departure had to be made because she had not found her north yet. She had to go further north. She then decided that she must walk to the North Cape.

She decided to walk the distance partly because she had no money, and partly because she wanted to do it the natural way, which was to use the body and the feet. Sofia started walking with little personal belongings, just a rucksack with a sleeping bag, a tent and a few clothes. She walked approximately 30 kilometres a day in six months, but had a few stops along the way. She slept outdoors but did also stay in private homes and had a few odd jobs here and there. She kept to the smaller roads and avoided the major roads and did not cross uninhabited areas and mountains alone. When she arrived at the North Cape in August it was cold and there was even a bit of snow, since fall and winter weather comes early. In fact, she wanted to leave and go south again immediately, because it was cold and wet and she longed for warm weather.

When departing from Oslo, the North Cape was nothing more than a point in an external and abstract geometric space. It was just a location, a point on a geometrical map. She knew from books she had read that the North Cape was like the end of the European world, far north. She had read about the fjords in the western parts and snow in the winters. What she learned after walking was embodied knowledge and understanding, because she had walked through the country herself. She gained these insights very slowly, because to walk is a very slow process of movement. She learned from what she saw and experienced in nature, and she learned from people she met.

Sofia wondered why other people travelled to the North Cape as tourists. She met many and asked them about it, but they could not tell her why, she said, because they talked only about all the other tourists, not their own authentic reasons, if they had any. Sofia wondered much about this, and came to a conclusion that many people travel to the North Cape because they view it as an object to collect. This is very different from how Sofia came to see the North Cape after she came there. When she arrived at the North Cape she felt that she had come to a very special place. Not because the place was special in itself, but because the arrival was meaningful in the context of her reasons to go and the long journey. It was special because of her experiences and learning over a long period of time. The truth of North Cape was not found in the gravel-coated flat landscape of North Cape, its magnificent view of the Arctic ocean or the myths of the midnight sun. She did not see much of this. Truth was not in the ground, as if truth is gold waiting to be uncovered by digging and mining. The gold was rather in the walking and in the internal journey that took not only place, but time. The meaning of the north for Sofia was in the travel method, the process of travelling and walking, not in the goal of the journey.

The Experience of Walking

Sofia told me that when she started walking, she was very depressed. She was in the middle of a life crisis, she told me. She had lost her north, and sought a new reference point for her life. While wandering she felt that the new power or energy came from herself through keeping the intention to walk. Through continuing to walk she felt she got her courage back. She also received much help from people she met while walking, in terms of a meal or a place to sleep overnight. The changes that took place in her came as a result of what she did herself, with herself, and in relation to her encounters with nature, with places, and the long and patient walking towards the north. Very slowly she experienced that she got better. Rather than becoming more tired and exhausted she felt that she gained more energy both mentally and physically. She did not know why she felt better. She only felt that she was on the right track, every day. If she met a problem, a solution would come. She did not, however, meet any serious problems, and people she met were always kind.

Sofia thought much about this. She thought her healing had much to do with walking and being bodily immersed in a natural environment much of the time, sensing the body's interaction with nature. The body is the best way to travel, she explained to me, since it makes a fine natural rhythm through movement. All other means of mobility are human creations, Sofia said. Cars, trains, bicycles, everything else, are human inventions. In our time people are very concerned about travelling faster and longer, the faster the better, but this is not natural, Sofia exclaimed. The body has a natural sense of movement and rhythm that takes outset in walking. And such travel takes time. Sofia said:

> To find myself, I had to use my own rhythm, and this is in walking. I walk inside myself to be inside myself, to look for myself, to use my natural rhythm. This is also walking outside of myself. I have to use the physical world because I am here on earth, and I have to use the ground, the water and other physical things that I can touch. To do that – to travel outside – I had to walk. That is a direct expression of my inside walking.

Walking is a particular form of mobility. Compared to modern transport technology, bodily movement mediates experiences of the external world, the world outside of the body, in a different way than what experiences that comes from travel by car, or by plane. Modern travel depends on transport technology in one sense or another, and the relationship between people and place are influenced by such technology. Sofia's journey, and the story of her journey, opens up for other experiences. When walking, one's experiences of nature, of places, and all living and non-living things, are directly sensed and perceived. One can also stop whenever one likes, one can listen, smell, and touch, and still be in the middle of one's own life whenever one likes, and wherever one is. There is one thing one cannot do when walking: one cannot cross a vast distance in a few hours. It takes time to wander.

Place and Placeless-ness

Time is a scarce resource for modern people. But what about place? Do we have place in our lives? Is there a scarcity of place? Do we feel placeless or that we live in a placeless world? As it has been argued in several ways, modern human beings have in a couple of hundred years crossed vast distances through technological development, made the world smaller and the ego bigger. Time and place have become resources to use and allocate. Most people take a plane when they want to visit the northern parts of the Nordic countries. The distance which Sofia spent months to cross, is done in three to four hours by plane. Time-space compression has shrinked the world. Everyday life in the modern world is not organized around bodily movement. It is organized on transport and communications technology. Holiday travel has gone through great changes in the twentieth century. Modern people are, to a great extent, travelling human beings, and modern societies are increasingly mobile ones (Lasch and Urry 1994, Bærenholdt et al. 2004). The question is no longer being, or being-in-the-world, but conquering as much space as possible in the shortest span of time. Being in place is also being in time, but technological development has structured everyday life to such a degree that it seems we have forgotten how to be in the world. It seems that it has been a modern project to go the furthest, and travel the fastest.

I will relate Sofia's experiences of walking to the need for sustainable and meaningful experiences of place. Place in this sense is not only a cognitive phenomenon, it refers to an experience of having a place and being place, in an emotional, embodied and cognitive sense. Place is basic for how human beings experience the world, and it is immediately more lived rather than known (Tuan 1977). It is a concept for feeling whole and fully human in a world of both human and more-than-human others, of relationship and reciprocity. Place is important because it is related to human self-understanding and to the understanding of the self's relation to the surrounding world. Being modern has to a great extent meant that humans have learnt to externalize the material world in opposition to the mind, the internal world (Irigaray 1985, Plumwood 2002). This is a typical example of dualist thinking in Western rationality. How are we to live and write relationships between self and world without dualist thinking? And, how is it possible to reconstruct community among humans, and common interests in the affirmation of the connectedness to place and nature, and simultaneously respect the right to diversity both for humans and the more-than-human world?

One inspiration for such reconstruction is to look again at Edward Relph's book *Place and Placelessness* (1976). He wrote that one of the aims of phenomenological research in geography should be to start a reawakening and a sense of wonder about the earth and its places, a sense of wonder about the world (ibid: 16). For Relph, geography is to have and to express a curiosity about and towards the world, and to learn to understand human *geographicality*, a term from the French philosopher Eric Dardel, who wrote that 'Geographical reality demands an involvement of the individual through his emotions, his habits, his body, that is so complete that he comes to forget it much as he comes to forget his own physiognomy' (Dardel 1952 cited in Relph 1981, 21).

Human relationships and belongings to place express identity, and personal identity will in many ways be identical to place identity. For many, place identity is also related to the presence or absence of good health or welfare. Relph says that 'to be human is to live in a world that is filled with significant places: to be human is to have and to know your place' (Relph 1976, 1). All places are experienced individually, and each individual stands in an individual relationship to his or her environment. In a phenomenological interpretation, the self and place reciprocates one another (Abram 1996). Place is not space, a homogenous void, but 'reveals itself as this vast and richly textured field in which we are corporeally immersed', a 'vibrant expanse structured both by a ground and a horizon' (Abram 1996, 216). Place reconciles space and time, as a living depth. This sense of place does not refer to location in space. Place does not necessarily have anything to do with particular locations. Even though place sometimes is understood as *locality* or geographical position, or *locale*, which is the physical setting or scene for social interaction (Cresswell 2004), the particular notion of place that Relph evokes refers not only to *sense of place*, the subjective and emotional attachment people have to place. It refers to the fact that being human is also being in and with place, or perhaps being and living place in a spatio-temporal sense (Birkeland 2005). This is an understanding of place where the temporal and the spatial is reconciled, and where the past is reconciled with the future (see also Chapter 3 by Mazzullo and Ingold and Chapter 17 by Benediktsson).

Relph exemplifies such a way of being place with Claude Levi-Strauss' narrative of his first journey to Latin-America, which was by ship. The experience of travel was according to Levi-Strauss nothing like an experience of placeless-ness, because the ship felt more like a home than being away from home (Relph 1976, 29). Places are culturally created, and any cartographic significance of place refers contingently to the experience and meaning of place. Nomadism, migration and mobility generally refer to different forms of localized experiences of place. Relph exemplifies this by the way of life of the Bororo-people in Brazil. The Bororo maintain a strong feeling of belonging to all the places they move to. Relph says that it is also a myth that human beings today who travel the most, and who are most mobile, experience homeless-ness and placeless-ness the strongest. Some experience a stronger identity or belonging to a place by being more mobile.

Geographer Anne Buttimer has also explored place and movement with particular reference to the dynamism of life-world (Buttimer 1976). She makes a very fluid conceptualization of place and argues for the necessity to think of place and time as connected. She treats life world as a whole new scheme for geographical thinking stressing rhythm and bodily processes. The vantage point of everyday life is place, Buttimer shows, and which is orchestrated through stabilizing and innovative forces, through rhythms of different scales (Buttimer 1976, 285). As I understand Buttimer, she also brings the non-conscious into the discussion as forces that set the human being's relationship to the world in motion. The living and sensing human body is the starting point with its 'rhythms or sound and silence, light and dark, smell, movement, and land use of the hours and days' (Buttimer 1976, 290).

These views on place and mobility stand in contrast to the long-standing view that place was not considered relevant in ethical or policy terms. Place was seen as static and apolitical, associated with border-makings and romanticized and idealized images of home, of the feminine, of the past, or with naïve conceptions of the local world. Now, place is seen as highly ethical and political, in many ways, and especially with reference to disourses on globalization and the reduction of biological and cultural diversity in its geographical sense, geodiversity. In this sense, Relph was writing against the placeless-ness of modern societies before many of the globalization critics. As an experiential term, placeless-ness signifies not feeling at home in a place, not knowing one's place. This represents a psychological and existential problem. Loss of place or feelings of placeless-ness express non-belonging or the experience of not having a place in the world. It expresses loss of a 'home' or place (Bachelard 1984; Relph 1976).

I think we have to see placeless-ness as a pathological condition, or a social disease, in modern societies. From a geographical perspective, we may interpret this in terms of absence or presence of authentic living (Relph 1976, 1981). Relph considered placeless-ness a double or paradoxical process. On the one side placeless-ness works as a non-planned destruction of distinct places. On the other side placeless-ness is a planned process of standardized landscapes which is based on insensitivity towards the subjective experience of place for human beings. Relph mentions the mass-images of places, for example travel destinations for mass tourists, as meanings of place which are neither individual nor group based experiences, and which communicates nothing but placeless-ness. The images of such place-less places are created by opinion-makers, provided ready-made for the people, disseminated through the mass-media and especially by advertising, Relph argues (Relph 1976, 58). Such places have the most superficial images a place can have, and the reason is that the meanings are not based on authentically experienced symbols, intentions, meaning, values and norms, but 'on glib and contrived stereotypes created arbitrarily and even synthetically' (Relph 1976, 58). This creates experiences of placeless-ness which should be reworked and transformed to more authentic experiences of place, Relph argues.

From Placeless Places to Authentic Place-worlds

While I agree that modern societies create placeless-ness, and recreate the conditions for placeless-ness, the focus should be on what is to be done about it. How may we create conditions for sustainable and authentic place-making? I would like to discuss the understanding of placeless-ness as a lack of relationship to ground, to nature and embodied nature. The word authenticity is in the centre here. According to Relph, authentic living is not achieved through 'right planning' or 'good design' of human environs. Solutions are not to be found in methods, but in radically changing our conceptions of self and the world. It is achieved by changing our relationships to place, our ways of doing and thinking.

In his second book, Relph developed his view of place, geographicality and wonder about the world to include the environment (Relph 1981). *Rational Landscapes and Humanistic Geography* is a critique of humanism and simultaneously a reconstruction of humanity in terms of a humility that denounce any idea of anthropocentrism and the primacy of mind and humanity over materiality and nature. With the term *environmental humility* Relph argues for a particular direction for a way of thinking and doing:

> ...that respects what there is in the world and seeks to protect it and even enhance it without denying the essential character or right to existence...Environmental humility is marked by a concern for the individuality of places, and this requires a careful and compassionate way of seeing that can grasp landscapes as subtle and changing, and as the expression of the efforts and hopes of the people who made them. (Relph 1981, 19)

Environmental humility is a political programme, not a recipe or methodological guide. It implies that we have to 'come to accept places, buildings, people and objects for what they are and as they are, not merely because of their resource potential or research significance' (Relph 1981, 20). The human being is not in the centre, but a part of a continuum of nature and culture where human beings influence and are influenced. It is to be in the middle of the world, within our place-world, not above, and not under, not dominating and not dominated, but living reciprocally and respectfully within place-worlds in constant movement and flux. Environmental humility is an attitude that:

> ... informs us that there are no clear boundaries to either human or natural processes, that nothing happens with absolute certainty and in isolation, that no systems are closed, that relationships rarely work in one direction only. (Relph 1981, 163)

I think that Relph's concept of environmental humility could play a very important role for reconsidering our concepts of self and place in a particular sense. His conception is sensitive to the interaction between human beings, culture and nature, an interaction that is engendered in the term place-making. It does not place the human being into a self-effacing subservience to nature. But it does place the human being with a particular responsibility for guardianship and protection. As Relph puts it: 'We have an obligation to look after people, creatures, plants, landscapes and everything else as they exist, simply because they exist' (Relph 1981, 164). The humility that Relph asks for, means that there is a need to care for, to take responsibility for our own immediate and surrounding place-worlds. It is a fundamental project for reconnection and restoration of placeless-ness.

This kind of humility is exactly what the Belgian-born French feminist philosopher Luce Irigaray calls for in her ethics of sexual difference, while providing a more nuanced understanding of placeless-ness as a masculinist, disembodied and abstract placeless-ness. In *An Ethics of Sexual Difference* she writes that our age is the age of sexual difference (Irigaray 1993). In her view, the making of place implies sexual difference in a particular way, since any concept of place is also a discourse on sexual difference. Sexual difference means for Luce Irigaray the difference between the two sexes with reference to embodied difference (Birkeland 2005). She argues that it is not possible to rethink sexual difference without simultaneously rethinking our understandings of place and time. Sexual difference could in fact represent our salvation if we thought it through, she argues:

> The transition to a new age requires a change in our perception and conception of space-time, the inhabiting of places, and of containers, or envelopes of identity. (Irigaray 1993, 7)

And:

> We must, therefore, reconsider the whole question of our conception of place, both in order to move on to another age of difference (each age of thought corresponds to a particular time of meditation on difference), and in order to construct an ethics of the passions. (Irigaray 1993, 11–12)

Her point of view is to speak from the position of being a female body and in terms of female genealogies with a language appropriate for women. Irigaray has often been misunderstood, when she has argued that the human being has always been written in the masculine form (Birkeland 2000, 2005). Her view is simply that the human being has been modelled after its masculine form, not both forms of humanity. The reason is that human beings have forgotten the meaning of sexual difference. The starting point for Irigaray's rethinking of place is the body. She takes the understanding of place much closer to the living body. The body is the most concrete place that exists and it is the very first place a human being senses. Every child senses the world as a female and feminine world through the mother. The mother is the world, she is the first place, and Irigaray stresses the need to give the relationship between place and woman a positive value for both men and women, but in different ways for the two sexes. The consequence is that it is not possible to think space and place without thinking sexual difference. Any discourse on place is simultaneously a discourse on sexual difference (Birkeland 2005).

Place-making, as I see it, is thus a continuous process of making place for, respecting and caring for difference and coming to accept places, buildings, people and objects for what they are and as they are, as Relph argued with reference to environmental humility. Place is not unitary or contained, and the boundaries between places are not closed, autonomous or impermeable. Place is potential place, revealing what it means to dwell. It is not a question of fixing the meaning of place. The central concern is not to have place, but to be place and to make place. There is sensitivity to the possibilities for another way of planning, building and living with and within place, based in sustainable concepts of place and self (Braidotti 1999). When I use the word sustainable, I treat it in a very loose sense. I am not using it in an instrumental sense of the word but to start thinking about things and people in a way where everything is related to everything and where everything affects everything.

When the relationship between people and land is unbalanced and 'sick', as we see in many industrialized areas undergoing post-industrialization these days, or where we see placeless places and inauthentic living, we need to direct our attention to those processes that constitute place and make place a tool for an ethics of place. As Mick Smith argues, place is not only a field of experience and way of life, but a tool for an ethics and politics of difference in the context of the development of placeless places (Smith 2002). An ethics of place in Mick Smith's view is a call to understand the human situation with respect to the natural world, as a practical sense and ecological habitus. Also, Val Plumwood argues for an ethics and politics of place (Plumwood 2002, 2006). This ethics should not only be theoretical and discursive, it should be heartfelt in order to care as well as think about what we as humans do to our others, who are both human and more-than-human. Place is implicated in human lives, values and discourses in complex ways. Reason alone is not enough for creating good places, and good places need good relations between human beings and their social and natural contexts. Place-making is then to create reciprocal and sustainable relations between human beings, society and nature, in a way which makes our 'ecological relationships visible and accountable', as Plumwood puts it (Plumwood 2006). Place becomes in this sense a community, perceived as such and meaningful for human beings, where the constitution of this community results from the agency of the human and the more-than-human worlds.

References

Abram, D. (1996), *The Spell of the Sensuous* (New York: Vintage).

Bachelard, G. (1994), *The Poetics of Space* (Boston: Beacon Press).

Bærenholdt, J.O., Haldrup, M., Larsen, J. and Urry, J. (2004), *Performing Tourist Places* (Aldershot: Ashgate).

Becker, E. and Jahn, T. (eds), *Sustainability and the Social Sciences*, (London: Zed Books).

Birkeland, I. (2000), 'Luce Irigaray: mors makt', in Neumann (ed.).

——— (2005), *Making Place, Making Self: Travel, Subjectivity and Sexual Difference* (Aldershot: Ashgate).

Braidotti, R. (1999), 'Towards Sustainable Subjectivity: A View from Feminist Philosophy', in Becker and Jahn (eds).

Buttimer, A. (1976), 'Grasping the Dynamism of Lifeworld', *Annals of the Association of American Geographers*, 66, 277–92.

Cresswell, T. (2004), *Place. A Short Introduction* (Oxford: Blackwell).

Irigaray, L. (1985), *Speculum of the Other Women* (Ithaca NY: Cornell University Press).

——— (1993), *An Ethics of Sexual Difference* (London: The Athlone Press).

Lasch, S. and Urry, J. (1994), *Economies of Signs and Space* (London: Sage).

Neumann, I.B. (ed.) (2000), *Maktens strateger* (Oslo: Pax).

Plumwood, V. (2002), *Environmental Culture: The Ecological Crisis of Reason* (London: Routledge).

——— (2006), 'Shadowplaces and the Politics of Dwelling', unpublished paper.

Relph, E. (1976), *Place and Placelessness* (London: Pion).

——— (1981), *Rational Landscapes and Humanistic Geography* (Totowa: Croom Helm).

Smith, M. (2002), *An Ethics of Place* (New York: SUNY Press).

Tuan, Y-F. (1977), *Space and Place: The Perspective of Experience* (London: Edward Arnold).

Chapter 5

Re-centring Periphery:
Negotiating Identities in Time and Space

Gry Paulgaard

Introduction

The vision of a new world order, replacing the solid, fixed and stable past with fluidity and mobility, a free unbounded space within a global universe, is frequently used as a diagnosis of contemporary society. As Kirsten Simonsen shows in Chapter 2, such a diagnosis leads to a basic *mis*understanding of the relationship between place and mobility. Place and identity are linked to the past stability in 'pre-modern societies', whereas the present situation, characterized by unfettered mobility, annihilates the meaning of place so that the old identity between people and places disappears (Massey 2005; Simonsen 2005; 2007). Such an understanding is deeply rooted in an evolutionary assumption functioning as the doxa for much of the discourse of the relation between local and global processes: '…the notion that we move from the local to the global as if to a higher stage of world history' (Friedman 2006, 123). We shall see that this notion is not only used to distinguish between different historical periods but also between different geographical areas; between the centre and the periphery. This particular hegemonic understanding divides the world both in time and space; space turns into time, geography into history (Massey 2005).

The chapter that follows will show how this hierarchical vision also works as a core understanding of people and places in Northern Norway. I shall use empirical examples from my field studies among young people living in coastal communities in Finnmark, the northernmost county in Norway. I shall also refer to public debates in newspapers and other media about Northern Norwegian uniqueness and identity. This will show how place still counts in people's construction of identity. Even though the present mobility conquers distance in many respects and might seem to reduce the local, regional or even national uniqueness of both people and places, the experience of difference as located with place does not disappear. When the main focus is placed on lived experience, it is possible to examine the relation between place and mobility by analysing how people handle the flow of cultural products in the identification of themselves and their localities.

The *Social* Construction of Identity

A basic misunderstanding of the relation between place and mobility affects the understanding of identity. From classical theorists to the present day, considerable energy has been expended in describing how mobility affects traditional forms of community and collective identity, offering the individual greater freedom to construct his or her own uniqueness almost on his or her own, disconnected from all kinds of 'social or cultural baggage': 'We are not what we are but what we can make of ourselves', as Anthony Giddens (1991, 75) has said, focusing on reflexive self-construction and individualization. All the same, this does not mean that anything goes, that any construction may be equally valid and verified by others. This is a crucial point: the importance of confirmation through the response of others, with regard to both possibilities and constraints in the construction of identities.

We do not start from scratch when we set out to create meaningful constructions: we do so within a universe of meaning that is already structured. Perhaps it is more correct to refer to reconstructions, rather than constructions (Kjørup 2003). The fact that people are embedded in social relations in different social situations within universes of meaning defines what is possible and not possible. The importance of others, of responses from others, has fundamental meaning for the individual potential to create oneself in terms of uniqueness or identity, as we have learned from Mead (1934) and others. We need others in order to carry out a reflexive self-formation. This seems to have been omitted by some theorists, who focus on the endless freedom people have in their creative self-formation. The importance of others – both 'real others' and 'virtual others' – is crucial to the construction, performance and enactment of identity in our time.

Massey (2005) uses the term 'the encounter with difference', describing space as the dimension of the social. She maintains that space presents us with the existence of others. Identities are created by means of experiencing similarity and differences. Encounters with differences and otherness are important to the perception of the self as distinctive in personal, social and cultural terms. When local identities are presented as uniform, intra-group differences are played down because distinctions perceived as even more different are being played up. In my research on young people in northern settings I have focused on identity construction, particularly on how collective identities based on place serve to play down other differences such as social class and ethnicity. This does not mean that such differences have lost their importance. I shall return to this later: the point here is to stress the relational and contextual importance in understanding how people create identities and boundaries, similarities and differences in an ongoing complexity of impulses, information and possibilities. This is one way of understanding the interplay between place and mobility, analysing the consumption of cultural strategies as part of an attempt to discover the logic in the apparent present chaos (Friedman 1996).

Placing Identity in Time and Space

If we take the vision of the new world order for granted, the great concern about roots and cultural identities in our time may seem ironic (Sahlins 1994; Savage 2005). However, ethnic and cultural fragmentations and modernistic homogenization are not two opposing views of what is happening in the world today. On the contrary, these are two constitutive trends of the global reality (Friedman 1996). Both these trends have to do with the existence of others and encounters with difference (Massey 2005). The intensive practice of identity is characterized as the hallmark of our present period (Friedman 1996; Sahlins 1994), and this also applies to the northern part of Norway.

There has been an increasing awareness of the meaning and content of collective cultural identity in this region over the past few years. This has inspired books, reports, articles and debates, not only among researchers but also in other contexts. Some years ago (31 January 1998), an article in *Nordlys*, the largest daily newspaper in Northern Norway, claimed in a two-page article with huge headlines that young people 'have to move to be modern'. The report was based on the myth of Northern Norwegians as naïve and natural, living among the fjords and the fish. The young people interviewed claimed that such one-sided media accounts might lead them to cease identifying with the place where they live:

> Northerners, as presented in the media, are old fishermen in sou'westers, who swear a lot and talk about cod and their mothers' fishcakes. Modern young people in Northern Norway don't identify with this image.

Young people do not identify with old fishermen: is that so peculiar, demanding huge headlines? I find the intensity and temperature of the debate far more interesting. I also find it interesting that over the years since this newspaper article was published there have been many intense debates in newspapers and on radio and TV, where myths about people and places in the north have either been rejected and criticized or defended and praised. Important cultural symbols established over generations have been defined as outdated. In 2003, there was practically a people's mandate for 'Modern Tales of the North' and there were a lot of seminars with almost exactly the same title, saying 'Modern Tales of the North – do they exist?'

The intense interest and emotional temperature of these debates show that myths concerning place are of great importance in the present situation. In the light of assertions about the disappearing identity between people and places, and because myths are distortions and simplifications of reality (they are not 'real', as opposed to *logos* as truth), this may seem somewhat remarkable. When young people, regardless of obvious exaggeration, bother to discuss the content of myths in a public debate and use this as a threat to move from the north, we see that such myths are embodied with power. Even though the myths about people and places in the north are defined as outdated, they are not without influence. Lévi-Strauss (1955) has claimed that myths create a shared community of meaning and therefore act as a basis of thought and action. Myths may be attractive or dreadful, convenient to identify with or to reject. If myths are repeated often enough they may function as realities in a cultural sense, as powerful symbolic material in the formation of collective identities, sometimes independently of individual preference and choice.

Core Contrasts in the Geography of Difference

The claim that young northerners have to move from the north in order to be modern is based on the understanding of an opposition between 'modern' and North. The contrast between 'modern' and 'outdated' combines with the contrast between South and North. This is not just a Norwegian or Northern Norwegian phenomenon: it has to do with the hegemonic understanding of the distinction between the civilized centre and the more backward and primitive periphery. Jonathan Friedman terms this temporalization of space a mistranslation of space into time, involving both primitivism and evolutionism: 'The difference between them lies in the respectively negative versus positive evaluation of this temporal relation, an imaginary continuum' (1996, 5). The construction of centre and periphery as asymmetrical counter-concepts has both positive and negative connotations, but the content of the dichotomies is the same, coding people on the northern periphery as less civilized, more outdated, wilder, more authentic and even more magical and natural than people who live in the centre.[1] At a symbolic level, the hegemonic understanding of the northern periphery functions as an encounter with difference and the north becomes fundamentally different, a negation of civilized life in more central areas.

During the 1970s, there was a culturalist movement in the northern part of Norway that challenged descriptions of the region as primitive and 'out of date'. Northern Norwegian culture was valued and idealized as a contrast to modern urban life and the social and cultural life in the south. As part of a cultural struggle to revalue and upgrade Northern Norwegian tradition and culture this image was presented in academic texts, as well as in films and theatre and by visual artists, singers and songwriters. Many of those who were important actors in this identity project – whom I shall term intellectual, well-educated elite – lived their lives in regional centres far from the places and the life-forms they praised. However, this political identity project played an important role in the cultural struggle to revalue traditional ways of life on the Northern Norwegian periphery, at the same time as devaluating ways of life in urban centres.

However, something has obviously happened somewhere along the line, since both young people and others today show an intense resistance to images that only one generation ago evoked some of the most powerful symbols of Northern Norwegian culture and uniqueness. The new voices in the debate about collective identity among northerners seem to have a political agenda, just like those of the 1970s, trying to change the frame of reference for the construction of a collective identity. The identity politics of the 1970s highlighted significant differences between North and South, giving the north a more positive evaluation, in contrast to the centre/periphery myths and centre-defined political opinions devaluing life on the periphery. Today, the political identity project seems to emphasize similarities between North and South, claiming that both the places and the people in the north are as modern as the modern centres in the south. This does not correspond with the dominant conceptual dichotomization between centre and periphery. According

---

1   For other examples and discussions of this phenomenon, see Chapter 18 by Kraft and Chapter 19 by Guneriussen.

to this dichotomization, the intense claim to be modern and the great demand for modern stories about the north is understandable.

During the 1970s, identity politics was linked with a positive evaluation of traditional ways of living in the north. Some critics have called this a glorification of primitivism, claiming that even though the evaluation of people from the north has changed, and northerners are not stigmatized as much as they were previously, the exotic myths have been reinforced and most people do not feel comfortable being associated with them. One crucial purpose of this chapter is to show that such myths are not easily deconstructed, even though they may not correspond to people's experiences in their daily life.

Re-centring Periphery in a Local Context

Regardless of whether or not one grows up in the north – that is, on the periphery – according to the core constructions, young people live their lives not 'out there' or 'up there' but 'here', as they see it. The relationship between the centre and the periphery is also linked to mentality and experience; using this as a basis may cause the picture to be viewed somewhat differently (Cohen 1982). This relates to the 'radius' or extended networks within which one finds oneself, mentally, and those with whom one is being contrasted.

The young boys and girls I have studied are growing up in places where there have been major changes within just one generation. Fishing, however, has been an important activity in this area for many generations.[2] Since the 1990s, there has been a decline in work opportunities in the fisheries and a growth in tourism, social services and service industries. Earlier, it was more common for young people in this coastal area to work in the fisheries or at other unskilled jobs where there was no demand for an extended education, but education has become more important over the past ten or twenty years.

Young people in a place called Honningsvåg go to school and they work for hotels and other companies such as Rica, Rimi and Spar, names of hotels and shops that are to be found nearly everywhere in Norway, owned by national and international companies. In their spare time they visit cafés, where they drink Coke and lots of different kinds of coffee and tea, just like people of their age do in other places. They listen to the same kind of music and they watch the same films and TV programmes as young people on other latitudes and they go online, where they have access to lots of information from the World Wide Web, via the Internet. What is peculiar about the changes in Honningsvåg is the fact that they are not peculiar at all: we find the same changes in other places, too, in Norway and in other countries.

While studying young people in different-sized northern places it has been interesting to note the perceived importance of appearing as up-to-date as possible, compared with the urban centre, when it comes to clothes, music, leisure activities, technological equipment and the experience of holidays abroad. Of course they

---

2   See Chapter 7 by Siri Gerrard, based on studies in Skarsvåg on the same island and in the same municipality as Honningsvåg.

know that they live on a periphery, compared to Oslo, London or Paris, but in their local and mental world they use every possibility of creating their own images and collective identities.

Young people living on the coast of Finnmark at a place called Båtsfjord make a distinction between themselves and young people living in Berlevåg, another place in the vicinity. Seen from the point of view of Båtsfjord, their own place is much larger and more modern than Berlevåg. The concept of 'modern' is used in the everyday language of these young people, in contrast to that of 'outdated'. By drawing a distinction between themselves and people from Berlevåg, the young people of Båtsfjord may perceive themselves as more modern.

Such contrasts are also relevant to young people from Honningsvåg, a place with around 3,000 inhabitants. From a local point of view, this place is experienced as distinctly different from, and far bigger than, smaller places nearby. By drawing such contrasts, Honningsvåg becomes a 'town' and a centre, an important element of the collective local self-perception on the part of the young boys and girls growing up in this 'town'. Young people from smaller places in the vicinity affirm this kind of attitude. Many of them have to go to high school in Honningsvåg because this level of education is not available where they live. Just as both young and older people from Honningsvåg buy clothes in bigger cities when they get the chance, young people from smaller places buy new clothes in Honningsvåg. They say that it is fun to bring home new clothes that no one else has. Thus, the young people of Honningsvåg receive confirmation that they live in a place with a rather more modern selection of clothes than those in smaller places.

Those from the smaller places also say that the young people in Honningsvåg are extremely engaged in being up-to-date and in fashion: even the boys are really dandy, they say. The boys and girls from Honningsvåg do not deny this: absolutely not, they confirm it. Such characterizations serve as confirmation of being in fashion and up-to-date with modern trends in urban centres, an important aspect in the construction of local identity. According to the hegemonic understanding of distinctions between centre and periphery this may seem peculiar, since trends and fashion are not associated with the northern periphery. The point here is that the use of signs and symbols from the more globalized world of fashion may be transformed into expressions of local uniqueness by the way in which people handle the flow of cultural products in the identification of themselves and their localities.

Those who come from smaller places in coastal areas of Finnmark say that people from Honningsvåg make jokes about the smaller places and the people who live there. The humorous characterizations are often unflattering so, for instance, people living in a place only seven kilometres from Honningsvåg are called 'de ville bak fjellet' (the wild ones behind the mountain). This characterization of wild may be understood as the people living there being less civilized than the people living in the centre, in Honningsvåg. According to the core understanding of the difference between centre and periphery, the primitive periphery is coded as belonging to a different age from that of the up-to-date centre. The young people in Honningsvåg live in the periphery in relation to national and international centres. In a local context, the situation is experienced otherwise. Here, the difference from and contrast to other young people coming from even smaller places plays an important part in the construction of local identity. The construction of local uniqueness and identity seems to be based on encounters with difference that differ from the distinction between North and South. Nevertheless, meanings produced in particular local contexts code the distinctions between the centre and the periphery and are almost equivalent to the core contrast between modern and outdated – but they assume a different significance.

In contrast to the 'new' cultural movement of the 1970s, which praised traditional life in the north, the young people I have studied seem to praise modern urban life in the centre. However, they do not necessarily have to move physically in order to perceive themselves as modern: slight shifts in established meanings, moving the boundary between the centre and the periphery from a national to a local level, offer the possibility of making important distinctions in the construction of collective identities.

Reinforcing Core Contrasts within the Field of Tourism

Tourism presupposes increased mobility and has been used as an example of how global processes can, in a sense, be pinned down in certain localities. Honningsvåg is one such locality, placed in the North Cape municipality, the northernmost municipality in Norway. North Cape is the main attraction for approximately 400,000–500,000 tourists who visit Finnmark each year. The total population of Finnmark is 70,000, so the number of tourists is quite large (Olsen 2003, Lyngnes and Viken 1997). An Italian explorer, Francesco Negri, is often described as the first tourist to North Cape. He was there in 1664 on an expedition to the northern regions. Several tourist magazines, brochures and books use a quotation from his diary in presenting this place:

> Here I am now at North Cape, at the far tip of Finnmarken – at the very end of the world. Here, where the world ends my curiosity ends too, and I return happily, if God wishes so.

As a product of the tourist industry, Honningsvåg as a place – and particularly North Cape – has been promoted as being on the extreme, the outermost marginal edge of civilization, at the very end of the world (see Chapter 4 by Birkeland). For the people living here, particularly for the young people growing up here, as we have seen, it is the *centre* of their world. Nevertheless, the exotic branding of the place and the people on the 'outmost marginal periphery' is used in the marketing and promotion of the place as a cultural product.

Many of the local youngsters work at North Cape, at nearby hotels and in other jobs within the tourist industry during their summer holidays. In the encounter with tourists, the consumption of cultural strategies is completely opposed to the strategy in other contexts. Instead of minimizing the difference between themselves and people from centres in the south, they accept the idea of being perceived as exotic northerners. Some of the young people working in a hotel wore clothes that looked more like colourful Sámi clothes than the Norwegian national dress ('bunad' in Norwegian). Most of those who wear such a uniform do not consider themselves to be Sámi: they are Norwegians, with Norwegian parents and grandparents, they say.

Sámi people in this area have been stigmatized since the beginning of the twentieth century, resulting in assimilation within Norwegian society. People in coastal Finnmark have grown up in places where the local culture has not been inscribed as Sámi but as local, as Finnmark or Norwegian culture (Paine 1957, Olsen 2003). North Cape municipality has appeared Norwegian for generations, except for the semi-nomadic reindeer herders who live there during the summer season. Thus, most of the people in this place are not familiar with being associated with 'emblematic Sámi symbols'.

Despite this, young people working in the tourist industry say it does not matter if the tourists think they are Sámi: 'We give the tourists what they want', they say. One of the young boys told me that he had 'a summer job as a Sámi', working in a Sámi camp among the reindeer, selling traditional Sámi handicraft souvenirs and letting the tourists take photos of him. The boy is a Sámi, but not in line with the stereotypes of indigenous people as the exotic 'Noble Savage'. The tourist industry presents the emblematic image of the Sámi people as a counter-concept to images of modern culture (Olsen 2003). Like the other young people from this place, the boy considers himself 'modern' and 'in fashion' in both style and activities; nor has he grown up with reindeers in a 'traditional Sámi family'. Working as a Sámi has meant that he has had to dress and act in accordance with the stereotypical images of Sámi culture, although these images do not reflect the way that most Sámi in Norway live today.

In other settings in a local context it is far more important to distinguish between the Sámi and the Norwegians, even though people say that Sámi ethnicity is not considered to be as negative as in earlier generations. The worldwide interest in indigenous peoples and the ethno-political mobilization of the Sámi people may have affected relations between different ethnic groups in this locality too. But it seems not to have had so strong an influence on the local perception of the place and those who live here. There has been a strong cultural and political revitalization among young people with a Sámi background in other places in Northern Norway. It has not taken place in this area. Here, the majority of the Sámi population is perceived as 'visitors', living on the island with their reindeers during the summer and leaving for Inner Finnmark during the autumn. Reindeers grazing in people's gardens often provoke conflicts between locals and the reindeer owners.

The branding of the place and the people within the field of tourism represents the local culture in accordance with the hierarchical understanding of the distinction between centre and periphery. This way of 'putting culture on display' represents an objectification of culture as a product and should not be mixed up with the way that people identify themselves and their localities (Olsen 2003). Nonetheless, practices within the field of tourism reinforce the myths, stereotypes and core contrasts that people are trying to deconstruct in other contexts.

Coping with Ambiguity and Ambivalence

According to the dominant discourses, the uniqueness of the locality as the marginal edge of civilization, the very end of the world, refers to something quite different from being up-to-date and *au fait* with central trends in fashion. Even though the young boys and girls insist on being modern, drawing contrasts with other young people on the local periphery, they themselves are also young people on the periphery when viewed in a national or international context. In the light of modern subcultures in Manhattan or a European metropolis, the young Honningsvågers' struggle to be more modern may be perceived as the result of extreme provincialism. They seem to be more Catholic than the Pope, a well-known phenomenon in the struggle to be accepted within an established group or category.

At home it is quite possible to be accepted and receive confirmation of the construction of the local identity as modern, but this construction does not have an enduring resilience far from home. One of the girls said to me: 'Hvis æ kommer til Oslo føle æ mæ ikke så motekledd. Æ føle at æ kommer herfra. Du e jo Finnmarking.' (If I come to Oslo, I don't feel so fashionable. I feel as though I come from here. Well, you're a Finnmarker.)

Through myths and practices, in encounters with 'virtual' others and 'real' others, people discover that the construction of a local identity is based on contradictions. People know it and show it, as a young boy said: 'Vi bor i et gammeldags avholl, men æ trives' (We live in old-fashioned hicksville, but I like it). With humour it is both tolerable and possible to comment on stereotypes, prejudices and climatic realities in the north: 'Vi har sne og vind om vinteren og tåke og vind om sommeren' (We get snow and wind in the winter and fog and wind in the summer) – a common expression of the climatic conditions in this place.

Because of its location at 71° north, there are in fact very few trees in this place. Those that exist are planted and need special care and protection; the trees are wrapped up during the winter. The locals used to say that the trees symbolize the resistance and strong-willed capacity for survival of the people who live here, as described in a local saying: 'Vi bit oss fast i hver en stein' (We've hung on by our teeth to every stone) – even though they also know that the local population statistics tell quite another story.

This wisecrack confirms the myth about the northerners' great sense of humour, moulded by rough, dramatic and spectacular landscapes (see Chapter 18 by Kraft). I should like to add that their sense and use of humour has been moulded by local practice and the experience of ambiguity growing up in what is both a centre and a periphery. Humour may serve as one cultural strategy for handling the ambiguity on which the construction of local uniqueness is based.

A group of amateur performers from Honningsvåg have made a living out of humoristic comment on stereotypes and myths concerning the north and northerners. When they first started, they were taken aback by the overwhelming popularity the show received, not only from the local public. One of them explained that they had not expected the great demand for this kind of show from a general audience. For several years they have been 'compelled' to produce new shows and spend many weekends on tour each year. The performers have also been criticized for swearing and using bad jokes, being 'over the top' when it comes to sexual taboos and the Sámi people. One of their shows had the expressive title 'Født bak lyset' (Born Behind the Light). Both the title and the content of the show play with stereotypes and myths about places and people on the Northern periphery, quite in line with the core understanding of the distinction between centre and periphery.

These actors are involved in branding and 'selling' exotic myths, just like the actors of a different sort within the tourist industry. But there is an important difference: these people are performing the stereotypes using humour. Humour is not to be 'taken for real', but taken seriously in another meaning. As such, humour may be considered as both a cultural strategy and a cultural product. Humorous texts about people in the north are selling because they are recognizable to the audience. By using humour it is possible to comment on and laugh at the dominant hierarchical representation of the distinction between centre and periphery. In this respect, the actors are negotiating identity in time and space by playing with the core understanding of centre and periphery, South and North, at the same time as they confirm them.

The ambiguity and ambivalence that are rooted in a contradictory universe of meaning and interpretation of practices are linked with place, with different constructions and opinions about the north. People are continually confronted with differences and otherness in their encounters with 'concrete others' and 'virtual others'. Experience of differences between people and places, myths and stereotypes, and interpretations of climate and geography in relation to 'the world outside' become embedded in people's consciousness.

Negotiating Identity in Time and Space

This chapter has shown how the core understanding of centre and periphery is handled in different contexts, in the public media, in encounters with locals and within the field of tourism. Viewed externally, people – especially young people – seem to become identical to people from other places with regard to activities and appearances in many situations. Nonetheless, examining the way in which people handle the flow of cultural products through practices in their everyday life, it is possible to discover how place has been made relevant in relation to other places in the 'near' and more 'distant' world. The negotiation of identity based on the evolutionary assumption dividing the world in both time and space illustrates how place and mobility go together in the present situation.

Different constructions or place myths are used for different purposes. The young people's rejection of the myth about people in the northern part of Norway and their 'desire to be modern' may be perceived as a reaction and a response to 'old' myths and images. The distinction between centre and periphery is reconstructed in different contexts, even though the 'radius' and extended networks that define the meaning of this difference may not be stable. The young people's rejection of exotic myths about northerners is not consistent in every situation, as we have seen in the field of tourism. Here the stereotypes and the dominant discourse are reinforced, representing easily recognizable images (Olsen 2003).

An understanding of the difference between centre and periphery is related to 'layers of other constructions' (Aspers 2002), such as the difference between the modern and the outdated, or the civilized and the natural or primitive. This shows how core constructions are composed within an established universe of meaning, and are therefore of greater durability and stability than others.

Local, regional and national identities may be understood as myths, a construction of community on a territorial basis. But when territorial identity – national, regional or local – is made relevant, in encounters, through people's attempts to experience and mark their difference from people in other places, such collective identity forms do exist. At the same time, these 'answers' may serve as an example of how meanings and an understanding of regional and local character define us within a collective sense of belonging, whether we choose this or not. Even though we speak about constructions, we are dealing with reconstructions, because we do not start from scratch in creating meaningful constructions; we are doing so within a previously-structured universe of meaning that defines what is acceptable and what is not.

Growing up in the North also involves 'growing into' fundamental constructions, contrasts and distinctions that are related to both symbolic and natural conditions; layers of meaning inherited over generations, despite all the global influences. Fundamental constructions of uniqueness are pursued, even though the relationship between the centre and the periphery may be adjustable and not absolutely stable. This shows that the meanings relating to the centre/periphery distinction form a kind of basic foundation for other constructions that is not easily deconstructed, even though the material that constitutes this construction may vary.

References

Aspers, P. (2002), 'Vad skulle det annars vara? Om socialkonstruktivism', *Sosiologi i dag*, 2, 23–39.

Borofsky, Robert (ed.) (1994), *Assessing Cultural Anthropology* (McGraw-Hill).

Cohen, A.P. (1982), *Belonging* (Manchester: Manchester University Press).

Friedman, J. (1996), *Cultural Identity and Global Process* (London: Sage).

Giddens, A. (1991), *Modernity and Self-Identity. Self and Society in the Late Modern Age* (Cambridge: Polity).

Kjørup, S. (2003), *Menneskevidenskaperne: Problemer og traditioner i humanioras videnskapsteori* (Frederiksberg: Roskilde Universitetsforlag).

Lévi-Strauss, C. (1955), 'The Structural Study of Myth', *Journal of American Folklore*, 68, 428–44.

Lyngnes, S. and Viken, A. (1997), *Samisk kultur og turisme på Nordkalotten* (Oslo: BI).

Massey, D. (2005), *For Space* (London: Sage).

Mead, G.H. (1934), *Mind, Self and Society. From the Standpoint of a Social Behaviorist* (Chicago and London: The University of Chicago Press).

Olsen, K. (2003), 'The Touristic Construction of the "Emblematic" Sámi', *Acta Borealia*, Nordic Journal of Circumpolar Societies, 20:1, 3–20.

Paine, R. (1957), Coast Lapp Society, *Tromsø museums skrifter*, Volume IV.

Sahlins, M. (1994), 'Goodbye to Tristes tropes; ethnography in the context of modern world history', in Borofsky (ed.).

Savage, M., Gay B. and Longhurst B. (2005), *Globalization and Belonging* (London: Sage).

Simonsen, K. (2001), 'Rum, sted, krop og køn', in Simonsen (ed.).

—— (ed.) (2001), *Praksis, Rum og Mobilitet: Socialgeografiske Bidrag* (Frederiksberg: Roskilde Universitetsforlag).

Chapter 6

Villages on the Move: From Places of Necessity to Places of Choice

Gestur Hovgaard and Sámal Matras Kristiansen

Introduction

The Faroe Islands constitute a small archipelago in the middle of the North Atlantic Ocean, situated between Iceland, Norway and Scotland. The land width is only 1,399 square kilometres, spread out across 17 inhabited islands and several islets. The archipelago, which has home rule within the Kingdom of Denmark, stretches 120 kilometres from north to south and 75 kilometres from east to west (Guttesen 1996). Although there may have been an earlier presence of Irish ecclesiastics, the Faroes are normally considered to have been first inhabited by Norse (Viking) settlers during the ninth century (Stumman Hansen and Sheehan 2006, Stummann Hansen 2003). Since the first days of settlement, rough seas and steep mountains have constituted quite clear divisions between the scattered village communities. These difficult terrains have considerably limited the mobility of the common people, and for the vast majority village life has for centuries constituted their sole life horizon.

With modernization, the close ties within the village communities have loosened on a continuous basis and they have opened up towards each other and the outside world. Over the past few decades, more and more villages and even islands have been connected by tunnel, bridge and road. The recent construction of two subsea tunnels means in practice that today more than 85 per cent of the 48,000 inhabitants live within approximately one hour's driving distance of each other. The infrastructural extensions over the past few decades have constituted a transitional state, which may have been extended by the fact that, increasingly, more people travel abroad for leisure, business and educational purposes. Even so, we maintain that the perception of the villages as something stable, something authentic, a particularly secure way of living, and entities in their own right still constitutes a strong source of both identity and policy formation. Seen from the perspective of technological and institutional modernization, these perceptions of village life may be rather romantic, and evidently do not correspond with the social and economic realities of today. What, then, are the realities of modern village life in the Faroes?

This chapter provides an analytical discussion of this difficult issue. Our point of departure is that villagers today increasingly commute between their everyday functionalities of home, work, leisure and shopping. In other words, village life is constituted by the fact that they are connected in what may be termed a 'village network'. First of all, our ambition is to grasp this new reality of 'village networks' (or 'island networks'), which in fact constitute an emerging Faroese Network Region. The term 'network region' is inspired by Castells' notion of the 'network city' (Castells 1996), though Castells' concept of the network city is here widened to include physical transport. We further argue that we are witnessing a transformation that constitutes not only increasing mobility but also reflexive settlement choices. The combination of mobility and reflexivity then becomes a central strategy for coping with modern life modes.

In order to outline the realities behind the 'network village', we shall first present the Faroese village from an analytical perspective as a historically changing reality that has always reacted to the forces of modernization, creating modernity variants of its own. Thereafter we shall outline a case study from a contemporary village municipality, together with ways in which villagers have coped with increasing mobility and reflexive settlement choices. Against this background, the article will end with a discussion concerning some of the perspectives of the Faroese Network Region.

The Faroese Village – Continuity and Change

The complexity of Faroese village development may be described by the analytical construction 'continuity within change' (Hovgaard 2001a; Hovgaard and Apostle 2002, 16–18). 'Continuity within change' signifies a different view of modernization from the classical one. In a classical perspective, modernity is perceived to be floating from the centres to the peripheries. Here, instead, modernization is perceived as a complicated mix among institutional orders at different levels, embedded in the social continuities of the village communities. From this perspective, Faroese villages have experienced three major transitions in modern times, all of them closely attached to changes in the international community and strongly affecting their social and economic cohesion (Hovgaard 2001a). During the late nineteenth century, the community-based farming economy, based on the rights and privileges of the big farmers, was transformed into a market-based and internationally-oriented fisheries economy (classical liberalism). A second transition took place during the aftermath of the international crisis of the 1930s and the subsequent Second World War. The fisheries remained the predominant economic activity, but the role of the public sector in managing economic and social affairs increased strongly and this became the central institution in development. The second phase, at least in its culmination, may be termed Fordist modernization. The third transition, due to its uncertain composition, is normally considered Post-Fordist (Hovgaard 2001a, 113ff). We are still in the midst of this transition, one in which not only are the economic structures changing fundamentally, but in which the basic condition of life has been formed by reflexive modernization (Beck et al. 1994; Hovgaard 2001a, 246ff).

Based on this analytical construction, the following subsections will provide a brief introduction to historical change in Faroese villages.

The traditional farmer's village

During the eighteenth century, there was only one 'town' in the Faroe Islands, the provincial capital and market town Tórshavn, with approximately 500 inhabitants. The remainder of the settlements consisted of farming villages: sets of farms typically clustered in valleys at the end of fjords and bays, with access to coastal fishing and whaling. The valleys provided the agricultural land and in the mountains surrounding the villages there were herds of sheep and wild birds. These villages were partially the hinterlands of the royal Danish export haven in Tórshavn, but they were also relatively autonomous and stable production entities in themselves, ruled by the local upper class, the farmers. Common people lived in smaller houses close to the farms that they served. The largest farming villages were also places where the priests lived, since the livelihood of the priests consisted mainly of the surplus from the land belonging to the parsonage.

Despite being stable farming communities, the villages were also poor, entirely dependent on agriculture, coastal fishing and bird, seal and whale hunting (Joensen 1987). During times of hardship, the poorest had to collect seaweed from the beaches in order to survive. Due to the difficult livelihood structures, the population did not increase for centuries (Guttesen 1969). Until the early nineteenth century, the population remained stable at approximately 4,000. The predominant production for export was wool products, mainly socks, which the locals sold to the Danish royal monopoly store in Tórshavn (Finnsson Johansen 1997).

The traditional farming village was not disembedded or dissociated from the international community, and the law and order of God and the King were represented locally. Nevertheless, it is fair to argue that the villages were primarily embedded within their own form of social organization, that is, localized norms and habits, and a distinct local hierarchy regulating economic, political and cultural spheres of life.

The modern fisheries village

Institutional changes in the mid-nineteenth century enforced a new era for the Faroese villages. A significant event normally mentioned in this respect was the abolition of the Danish royal monopoly in 1856 (Apostle et al. 2002, 28–9). The rigid social bonds of the traditional village, including the rights and the privileges of the farmers, became defunct. The ordinary men became fishermen, while the women became fish-workers and around the turn of the century long-distance fisheries finally took over as the main occupation (Hovgaard 2001a, 150; Guttesen 1996, 7).

Capitalist entrepreneurs (ship-owners and businessmen) became the new upper class, while the farmers slowly but steadily drifted towards a more peripheral status in the economy. What evolved in this period was the rise of village capitalism, that is, one in which local entrepreneurs invested in sea-going fishing vessels (mainly sailing ships, that is, smacks/sloops) that exploited the waters around the Faroes, Iceland and – later on – even Greenland.

The social structure of the villages changed dramatically, and the village structure itself became more diverse as those villages with the most favourable geographical conditions for harbours started to grow. The first villages to develop in this direction were Vágur and Tvøroyri on the island of Suðuroy, and people from all over the islands moved to these villages to earn an income. Later, Klaksvík, Runavík, Fuglafjørður and Vestmanna also joined this category (Holm and Mortensen 2002).

The new fishing villages naturally became the economic locomotives of the Faroes and the whole structure of the Faroese villages was in fact stricken by a phenomenon rather uncommon in the traditional village: mobility, in the form of seasonal and permanent mobility.

Increased mobility is primarily exemplified by the new fisheries villages, which were also home to the medium and long-distance fisheries in a system of local-international capitalism (Hovgaard 2001). The traditional villages maintained a strong position, however, since they remained important for agricultural production and furthermore supplied labour to the new fisheries villages. They more or less managed to combine the best of both worlds – farming and fishing – forming specialized village life modes (Bærenholdt 1991, 345).

Another example of increased mobility, at least during the first period of transition, is illustrated by the fact that much of the old coastal activity was transformed aboard the smacks, for instance by using the vessels to carry the old coastal boats, which then could operate from the Icelandic coast. There were even women who sailed to the Icelandic coast, where the crew settled ashore for the season (Haldrup and Hoydal 1994, 110).

In addition, during the late eighteenth century and the whole of the nineteenth century, a new type of village, the settlement village (Niðursetubygdir), emerged (Finnsson Johansen 1997). People simply 'moved out' of their old villages and settled in new places – typically near their old homes – where they could settle their own plots. By the early part of the twentieth century, the number of villages had increased from the original 85 medieval villages (Markatalsbygdir) to more than 130 villages.

The development of the modern fishing village is truly a story of technological and institutional innovation, a story for which we do not have space for full justification here. Up until the 1960s, the Faroese fisheries economy was characterized by old technology and work combinations. A renewal of the fisheries fleet was started in the late 1950s, based on modern technology in the form of coastal boats and longliners, that is, a technology that favoured village development. But from the 1960s, and in particular from the mid-1970s onwards, the central strategic developmental component became mass production in the form of highly modern processing plants and freezing trawlers. On one hand, policy, finance and commerce formed a corporate system (Mørkøre 1991) that promoted growth on Fordist principles. On the other hand, strong territorial (village) interests that were able to attract investment in modern ships, plants and infrastructure (in Faroese called _bygdamenning_ – village development) balanced this modernization. In this way, many fisheries towns and villages were able to attract their share of societal advances.

The problem was that the institutional system was too weak and became dominated by a competitive race between territorial and sectoral interests; for only as long as there were financial supplies available and fish to exploit was it possible to maintain the modernization race. This was the logic of 'destructive competition', resulting in over-exploitation and over-investment, and the Faroese economy basically collapsed at the beginning of the 1990s (Hovgaard 2001a, 157ff).

The Emergence of a New Village Type – the Networking Villages

The fisheries villages have maintained their strong position in the Faroese economy and society up until the present time. The process of modernization has been mainly based on technological progress, however, and with the collapse in the early 1990s the fishing industry became centralized and rationalized, leaving many villages without a basis for existence. In addition, technological modernization normally intersects heavily with institutional change, change that until recently has been beyond the agenda of Faroese village development. This lack of readiness has clearly locked several of them into peripheral positions.

One of the most important changes that has already affected most Faroese villages for quite a long period of time is the changing youth culture. For most young people of today, working in the fish processing industry is only one of many options, and certainly not an attractive one (Holm and Mortensen 2006, 22, 27).

In general, the life horizon of young people has turned towards education, creative businesses and the service sector (Biskopstø and Mortensen 2004). For most youngsters this orientation involves out-migration, not only from their home village but even away from the Faroe Islands altogether (see also Chapter 5 by Paulgaard). Traditional craftsmanship and fishing is still a quite popular occupation among Faroese boys, and there may still be opportunities to stay on in the villages. However, the girls are clearly oriented towards the service sector, and even more so than the boys towards higher education (Biskopstø and Mortensen 2004, 24–6). From a village perspective, which does not normally imply a strong attachment towards services and higher educated people, the distorted gender variation is already a societal topic in some places and is reaching an inappropriate level in others.

In a paradoxical way it is the policy of village development, so heavily criticized for causing the economic collapse in the early 1990s, that has initiated the regionalization of most of the Faroes. The continuous interconnection of villages and islands has created the necessary geographies for what might be described as 'mobile modernity': a specific variant of Faroese reflexive modernity in which the most significant materialization is (auto)mobility. Everyday mobility is an option for most Faroese today. With the recent subsea tunnels – one opened in 2002 to Vágar in the west and the other in 2006 to Borðoy in the east – only 15 per cent of the Faroese population today lives outside the 'mainland'.

The fact that 85 per cent of inhabitants are now able to choose between everyday places means that the labour market that supplies the Faroese economy is no longer local but regional. Instead of being conceived as 'rural areas', the bulk of Faroese villages might rather be seen as suburban spaces that form part of a larger, coherent network region.

Increased mobility is illustrated by the fact that there was a 64.3 per cent rise in automobile traffic during the period from 1997 to 2006 (Hagstova Føroya 1). Furthermore, during the period between 1994 and 2006, traffic through the tunnel connecting the capital region and the northern regions rose by 122.4 per cent (from 2,470 to 5,494 vehicles on average per day – Hagstova Føroya 2, 550); use of the tunnel leading further north to Eysturoy rose by 182.1 per cent (from 1,248 to 3,520 vehicles on average per day – Hagstova Føroya 2, 610); and traffic on the road to the airport in Vágar rose by 100.3 per cent (from 1,217 to 2,438 vehicles on average per day – Hagstova Føroya 2, 440). The airport region was connected to the mainland by a subsea tunnel in December 2002 and, mainly because of this, traffic rose by staggering 27.2 per cent in 2003, compared to the year before (from 1,656 to 2,106 vehicles on average per day – Hagstova Føroya 2, 440).

The idea of the Faroes as a single functional entity is not only an academic idea: it is also taking shape in policy discourses. Firstly, a futuristic suburb in the northern outskirts of the capital Tórshavn is currently planned, which will serve most of the Faroes (á Dul Jacobsen 2006). Secondly, the Faroese government has proposed, in a vision report on Faroese infrastructure (Visjón 2015, Infrakervið) that the Faroe Islands should be perceived as 'one large city' by 2015:

> Today, 86 per cent of all Faroese are connected to the capital by road, and with good travelling opportunities all the Faroes could be counted as one centre. The goal is that the travel time to Tórshavn be comparable to travel within large cities. (…) Compared to the rest of the world the entire Faroes are a periphery and it is therefore necessary to have at least one place that can provide the same offers, services and supplies as the rest of the world. It is necessary to create a city spirit in order to gain and maintain a skilled workforce in the Faroes (…) Travelling to Tórshavn should last at most one hour where there is a connection to the mainland.

This signifies that the Faroes are now in the midst of a structural change that is finally abandoning 'the village' as the cornerstone of societal life, and instead should be perceived as interconnected, in the form of 'network villages' constituting a network region.

Gradually, the villages are no longer places of necessity but have become places of choice. Today you can live in one place, work in another place, buy commodities in a third, pursue leisure activities in a fourth and go to church in a fifth. This fact – constituted by high mobility – is a fundamental characteristic of the modern village in the Faroe Islands.

This fact poses many difficulties for those municipalities that have either not understood this functional shift or are unlucky enough to be situated at the outset of or beyond the main arteries of the network region, the highways.

Village Residence in an Era of Mobility

Those villages situated at or in proximity to the highways are finding new roles in a new societal context (Matras Kristiansen 2006). Several village municipalities have realized that they are competing with the towns in providing much cheaper housing and plenty of housing plots, supplemented by a guarantee of childcare and care for the elderly. One of the municipalities that has managed this functional shift quite well is the municipality of Gøta on the eastern shore of Eysturoy.

Gøta municipality has transformed itself from being a primarily fishing industry municipality to becoming, partially, a commuting municipality. There are about 530 jobs within the municipality, but only about 354 of these are occupied by people actually living in Gøta municipality (Gøtu kommuna, forthcoming). The workforce in Gøta comprises about 550 people, which means that approximately 200 people – more than 35 per cent of the workforce – commute to some other municipality. Conversely, this means that a similar number of people commute to work to Gøta from other municipalities. Mobility thus creates an increased supply of job opportunities for people living in Gøta. But the fact that the villages can no longer be perceived as production entities also has consequences for Gøta. These are numerous (Matras Kristiansen 2005), but in the following sections we shall outline only three particularly important themes: professionalizing childcare, coping with mobility and re-embedding identity.

Professionalizing childcare

Mobility as a coping strategy has a large range of everyday practical implications. The increased commuting demands time resources, but this also means that the functions within the residential community have changed. Not only do people commute more and farther than before, there are also more people in the labour market than before. Between February 1990 and February 2007 the overall labour force grew by 4.8 per cent. Female growth during the same period was 10.5 per cent, while male growth was less than 0.5 per cent (Hagstova Føroya 3). Thus, the female labour force has grown significantly and now accounts for almost 47 per cent of the entire workforce. The increasing mobility probably explains part of this growth, since mobility has increased the opportunity of getting a job matching skills and interests, which may have contributed to bringing women into the labour market. Yet what is important at this point is that the growing female labour force has diminished the time available for housekeeping and childcare. In past decades, young mothers could depend on grandmothers helping out, but today the grandmothers are on the labour market as well. This is one core reason why the Faroese rural areas have experienced a large increase in the professionalization of childcare over the past decade.

Until the crisis during the 1990s, the welfare sector was not really a concern for the municipalities. Social issues were mainly dealt with by the government. Social services used to be very labour market oriented, but over the past few decades most municipalities have realized that providing social services, such as childcare and care for the elderly, is an important factor relating to citizen satisfaction and even a competitive parameter in terms of tempting new settlers to move in.

A former member of the board of the municipality of Gøta, Andrias Petersen, now a member of the National Parliament, explains the recent policies of the municipality in this way:

> It must be a guarantee that – if you live in Gøta – you will have your children taken care of. We have also done an incredible amount to the school. Now there is only a single-stream school up to seventh grade, and the aim has been to have a school with a good reputation, where the children thrive and learn something and the physical framework is good. Because this is also what modern families are looking for. It is one of the criteria... The criteria for settling, what are they? Childcare, plots of land, the school and then the environment for the children to grow up in. That is what people are looking for when they are settling. (Matras Kristiansen 2006, 107–8)

The families settling in rural municipalities are – to use Petersen's words – 'modern families'. Municipal planning needs to understand the demands of modern families, but it also needs to be understood that this is a result of a shift in the role of the Faroese village. Mobility not only interconnects distant localities, it also broadens the everyday spatiality of the people. Mobility becomes part of people's structural reflexivity as these structures are geographically extended (Drewes Nielsen 2005). Residence has become a mobile practice, not just a territorial practice, as it would normally be conceived (Matras Kristiansen 2006, 108).

Coping with mobility

This means that municipalities have to accept that people living within their boundaries live there on only a partial basis. Naturally it is impossible to talk about 'home' if this is not a place where relevant production spaces may be reached within an acceptable amount of time, but with access to mobility capital (Urry 2000; Urry 2004) such as the automobile, the individual is in fact a regional citizen, not just a community citizen. Mobility is thus a strategic means of being able to change jobs and other networks without moving home (Båtevik et al. 2004, 140).

Conversely, this means that people with scarce mobility capital are less able to choose between spaces, whether this is a matter of work, leisure, education or shopping. While the rural areas of the Faroes are lucrative for mobile people, they have become difficult places to live in for people who are immobile, since the scarcity of mobility capital also affects their network capital, that is, their ability to fulfil their social obligations and social needs, as described by Urry and Larsen in Chapter 8. The increased spatial competition means that services have become centralized and many communities have lost facilities such as shops. In Syðrugøta, one of the villages in the municipality of Gøta, there is not a single shop left. Syðrugøta has been transformed into what might be called a 'sleeping village' (Finnsson and Matras Kristiansen 2006, 24; Matras Kristiansen 2006, 110). This does not bother the mobile settlers, since they are used to driving into the centres, but for children, poor people and elderly people living in Syðrugøta it has become a problem because they are dependent on relatives to transport them between everyday functions that have become increasingly segregated. In some areas public transport is fairly good during the day, but in many villages it is difficult to live there without daily access to a car. Thus, to immobile people, mobility can also cause 'unfreedom' (see Freudendahl-Petersen 2005 for a discussion of this topic). It could be said that for people living in the rural areas of the Faroes, 'mobility versus immobility' has become one of the major social fault-lines (Matras Kristiansen 2006, 109).

Mobility is not only a way of coping with a territorially-based everyday existence. Mobility has become a central part of everyday life. What has emerged is nothing less than an entire 'mobile culture' or 'mobile life mode' (see also Bærenholdt 1998, 205 for a slightly different discussion of local and mobile life modes), which has replaced the traditional, localized village life modes (as described in for example, Bærenholdt 1991).

Re-embedding identity

Even though the Faroese villages are increasingly interconnected functionally, they still have a strong identity factor for many young settlers. Many – or most – youngsters move to the capital Tórshavn or abroad, but there are still large numbers of people settling in the rural areas situated in the proximity of the towns. Even though total strangers do move to some villages, the bulk of the settlers are in some way related to the municipality where they have chosen to settle. Such relationships may be family-related, or relationships with mates or peers.

In the Faroes the dominant residential preference is to own one's own house, a fact that stems both from tradition and from a long array of financial advantages pertaining to a house that is owned. Even if people buy or build their own home, some of them still feel it is important to live in their childhood community. Even though the geographical range of job-opportunities has widened, the villages' sense of continuity remains. Even though people living in the rural areas on the mainland share a common mobile culture, there are still some preferences that matter. It is important to feel at home, home meaning not just the space within the physical formation of 'a house'. Home is also the house surroundings, the view over the landscape and the 'villagescape' (from the Faroese *bygdarlag*, related to *landslag* or 'landscape').

There is a huge difference between 'not being able to leave' and 'being able to leave, but wanting to stay'. The second notion is more reflexive than the first one and living 'in the country' is not necessarily, therefore, a practice descending from an inherited agricultural identity. So Faroese villages should no longer be regarded as 'places of necessity' but as 'places of choice' (Hovgaard 2001, 246ff). Because of increased (auto)mobility, living in the country is now merely a spatially different way of living in a suburb.

The Faroese Network Region – a Region of Mobility

The question is not so much whether the Faroes are becoming re-embedded as a network region, but rather how this new region will be formed. There seem to be three possible outcomes: firstly, a multi-polar region with several centres; secondly, a region with two dominant nodes (Tórshavn and Skálafjørður); or, thirdly, a region where rurality is entirely subordinated to the dominant node, which realistically can only be the capital Tórshavn.

The outcome depends on several factors. One important factor is that politicians outside the capital will be eager to sustain the centrality of their home areas and municipalities, as the Faroese electorate is very locally founded. Yet the most important factor will be infrastructural development in the immediate future. At the moment, actors based in the Skálafjørð region in the north-east of Tórshavn plan to construct a giant subsea tunnel under the fjord at Tangafjørð, which currently separates the Skálafjørð area from the Tórshavn area (see Figure 6.1). This giant investment will shorten the route from the Skálafjørð area and other areas in the north-east of the Faroes with a transport time of more than 30–40 kilometres, or approximately half an hour (which is significant in Faroese terms). The distance from the Skálafjørð region to Tórshavn will then be only 7–15 kilometres.

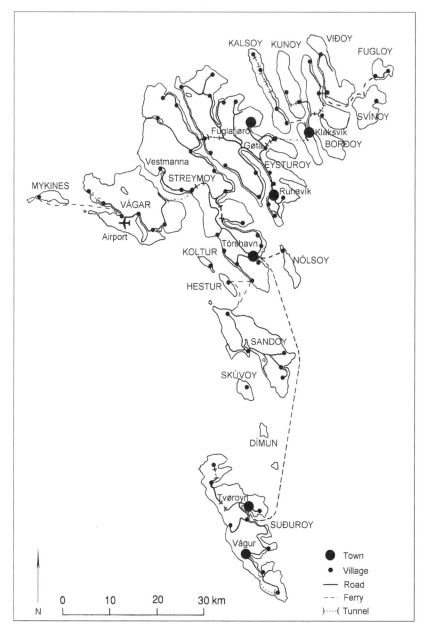

Figure 6.1   Map of the Faroes

Source: by Jørgen Ole Bærenholdt.

Population growth outside the towns is heavily concentrated in those rural areas that are favourably located along the highways connecting the towns (Matras Kristiansen 2004; Matras Kristiansen 2005, 153; Finnsson and Matras Kristiansen 2006, 22).

When or if the Skálafjørð tunnel becomes a reality, areas in the north-east of the Faroes will probably experience an increasing growth in transport and settlement. The main growth will be concentrated along the highway that links the capital Tórshavn to the south with the Skálafjørð region and Eastern Eysturoy in the middle and the Klaksvík region to the north. The centrality of the Skálafjørð area in the economic development of the Faroes was indicated years ago and with Tórshavn and Klaksvík in the south and north, respectively, this will constitute the 'growth axis' or 'central region' of the Faroes in the future. This hypothesis was stated years ago and still seems to be the most likely one (Apostle et al. 2002; Guttesen 1996, 64–5). This means that the Faroes community can anticipate an increasing population density on its south-eastern shores and a less dense population in all other regions.

Another dimension in the future settlement structure of the Faroes will be the ability to attract the considerable number of young people moving abroad. It has already been proposed that a housing shortage in the central towns has caused the population to stagnate over the past few years, but unfortunately no figures exist for this. One figure that illustrates the shortage of housing is the fact that there are currently (Spring 2007) more than 1,700 citizens in the capital, Tórshavn, who are in need of housing: a considerable number, almost 10 per cent of the entire population of Tórshavn municipality. Furthermore, there are Faroese living abroad or in rural areas who need housing in Tórshavn. These are seldom included in the statistics for Tórshavn municipality. There are several reasons other than the lack of appropriate housing that are contributing to people not moving back to the Faroes, but that remains beyond the scope of this article. One of these problems relates to the fact stated earlier in this chapter that it can be difficult to recruit skilled labour for the fishing industries (Hovgaard and West 2003). What is clear is that many people have to choose between buying into or renting from the hugely expensive private housing market in the centre of Tórshavn or, alternatively, moving to municipalities in nearby areas, which then take over the function of suburbs. One of the growth factors in the rural areas of the Faroes, therefore, is the fact of Tórshavn pushing out sections of its population and potential population. The motor of the Faroese Network Region is the push-pull dynamic between the centres and the peripheries that is mainly regulated by housing prices. It may be argued that this dynamic poses a threat to the continuities (in change) that have kept the villages strong over the past two centuries.

One obvious problem is that villages risk becoming second-class places that are subordinated to the towns. The quotation earlier in this chapter from the government vision report on the Faroese infrastructure certainly indicates such a rationale. Whether it will remain an attractive proposition for people who have experienced urban life in, for example, Copenhagen, Aberdeen, Reykjavík or London to move to the Faroese ruralities is a highly controversial question. If potential immigrants are not attracted to the rural areas and if the housing-market in Tórshavn becomes increasingly hostile then the Faroes will have nothing short of a national problem, since this may result in people postponing their move or giving up the idea of moving back to the Faroes entirely.

The Faroese Network Region has existed for several years, though it has not really been conceptualized until recently, and it is now establishing an increasing presence.

This has also peripheralized the discourse on regional development (*økismenning*) from below, which was promoted during the 1990s (for example, Hovgaard 2001b). In addition, the desire for public spending to drive the regional economies in the peripheries and semi-peripheries has proven out of time with political will and the rationale has therefore turned towards a full-scale integration of the Faroese village economies.

The question remains whether the Faroese villages will sustain their popularity as settlement areas. As long as the housing shortage and the ensuing sky-rocketing of housing prices in the towns continues, people who cannot enter the expensive housing market in the towns are being forced to move to the villages. The dynamic of the centres is pushing people out into the rural areas at the same time as it is pulling in those who can afford it. If the housing market should become better at satisfying demands, either through a larger supply of flats or plots or by a national economic recession, which would lower housing prices considerably, like the recession in the 1990s, then the question remains whether village people will move to the towns into which they are already functionally integrated. Why keep on commuting every single day, if it proves possible to move to a place of work where the broadest job opportunities are located?

References

Aarsæther, N. and Bærenholdt, J.O. (eds) (1998), *Coping Strategies in the North*, INS 1998: 303 (Copenhagen: MOST and Nordic Council of Ministers).

Apostle, R., Holm, D., Hovgaard, G., Waag Høgnesen, Ó. and Mortensen, Bj. (2002), *The Restructuration of the Faroese Economy* (Frederiksberg: Samfundslitteratur).

Båtevik, F.O., Mattland Olsen, G. and Vartdal, B. (2004), *Jakta på det regionale mennesket* (in Berg et al. 2004).

Bærenholdt, J.O. (1991), *Bygdeliv* (Roskilde: Forskningsrapport nr. 78, Institut for Geografi, Samfundsanalyse og Datalogi, Roskilde Universitetscenter).

—— (1998), 'Locals versus Mobiles' in Aarsæther and Bærenholdt (eds).

Beck, U., Giddens, A. and Lash, S. (1994), *Reflexive Modernization. Politics, Tradition and Aesthetics in the Modern Social Order* (Cambridge: Polity Press).

Berg, N.G., Dale, B., Lysgård, H.Kj. and Løfgren, A. (2004), *Mennesker, steder og regionale endringer* (Trondheim: Tapir Akademisk Forlag).

Biskopstø, O. and Mortensen, B. (2004), *Ung, Útbúgving og Fiskivinna* (Klaksvík; Vágur: Arbeiðsrit 9, Granskingardepilin fyri Økismenning).

Castells, M. (1996), *The Rise of the Network Society* (Cambridge, Mass.; Oxford: Blackwell Publishers).

Drewes Nielsen, L. (2005), 'Reflexive Mobility – A Critical and Action Oriented Perspective on Transport Research', in Thomsen et al. (eds).

Dul Jacobsen á, J. (2006), *Stóratjørn* (Tórshavn; Reykjavík: Selmar Nielsen Arkitektar; Arkís).

Finnsson, B. and Matras Kristiansen, S. (2006), *Bygdin, Fyrr, nú og í Framtíðini* (Vágur: Námsrit 7, Granskingardepilin fyri Økismenning).

Finnsson Johansen, S.T. (1997), *Fra 'hose til fiskefilet' – et studie i Færøernes teknoøkonomiske historie* (Roskilde: Arbejdspapir nr. 131, Institut for geografi og internationale udviklingsstudier, Roskilde Universitetscenter).

Freudendahl-Petersen, M. (2005), 'Structural Stories, Mobility and (Un)freedom', in Thomsen et al. (eds).

Guttesen, R. (1969), *Færøernes interne og externe migrationer 1956–1965 – Med kort beskrivelse af samfunds – og befolkningsudviklingen* (København: Afløsningsopgave i befolkningsgeografi, Københavns Universitet).

—— (1996), *Topografisk atlas Færøerne* (København: Det Kongelige Danske Geografiske Selskab).

Gøtu Kommuna (forthcoming), *Býarskipan, Gøtu kommunu 2007–2027* (Gøtu Kommuna).

Hagstova Føroya 1, *Koyring á Landsvegnum í Mió. Kilometrum Skift á Økir* (http://www.hagstovan.fo; pageid=33,162497).

—— 2, *Ársmiðalferðsla og Vísital Fyri Ársmiðalferðslu um Samdøgri* (http://www.hagstovan.fo; pageid=33,162501).

—— 3, *Løntakarar Skift á Vinnugrein, Kyn og Mánaðir* (http://www.hagstovan.fo: pageid=33,182512).

Haldrup, M. and Hoydal, H. (1994), *Håb i krise – Om muligheden af et menneskeligt fremskridt på Færøerne* (Roskilde: Forskningsrapport nr. 99. Publikationer fra Institut for Geografi og Internationale Udviklingsstudier, Roskilde Universitetscenter).

Holm, D. (2004), *Fjarferðing í Føroyum* (Vágur: Arbeiðsrit 8, Granskingardepilin fyri Økismenning).

Holm, D. and Mortensen, B. (2002), 'Economic Life in the Periphery. A Comparative Study of Economic Development in two Towns on Suðuroy', in Apostle et al.

—— (2006), *Arbeiðsmegin á Føroysku Flaka- og Fiskavirkjunum í Framtíðini* (Vágur: Arbeiðsrit 11, Granskingardepilin fyri Økismenning).

Hovgaard, G. (2001a), *Globalisation, Embeddedness and Local Coping Strategies. A comparative and qualitative study of local dynamics of contemporary change.* (Roskilde: Department of Social Sciences, Roskilde University).

—— (2001b), *Økismenning á Beina Kós* (Klaksvík: Arbeiðsrit 1, Granskingardepilin fyri Økismenning).

Hovgaard, G. and Apostle, R. (2002), 'Case Studies of Socio-economic Change: Analytical Concerns', in Apostle et al.

Hovgaard, G. and West, H. (2003), *Rekrutering av Høgt Útbúnari Arbeiðsmegi til Fiskivinnunar* (Klaksvík: Arbeiðsrit 4, Granskingardepilin fyri Økismenning).

Joensen, J.P. (1987), *Fólk og Mentan* (Tórshavn: Føroyar Skúlabókagrunnur).

Lewis-Simpson, S.M. (ed.) (2003), *Vínland Revisited: The Norse World at the Turn of the First Millennium*, Selected Papers from the Viking Millennium International Symposium, 15–24 September 2000, Newfoundland and Labrador (St. John's, Newfoundland: Historic Sites Association of Newfoundland and Labrador).

Matras Kristiansen, S. (2004), *A New Role for the Faeroese Village?* (Klaksvík: NOLD7 working paper).

—— (2005), *Forestillingen om bostedet* (Roskilde: Non-published MA thesis in Geography and Social Policy, Roskilde University).

—— (2006), 'Village Dwelling. Understanding young peoples' reasons for settling in Faroese villages', *Fróðskaparrit*, 54, bók 2006, 96–113.

Mørkøre, J. (1991), *Et korporativt forvaltningsregimes sammenbrud – Erfaringer fra det Færøske fiskeri i nationalt farvand* (Visby: Paper for the Nordic Seminar on Fisheries Issues, unpublished).

Stummann Hansen, S. (2003), 'Scandinavian Building Customs in the Viking Age – The North Atlantic Perspective', in Lewis-Simpson (ed.).

Stummann Hansen, S. and Sheehan, J. (2006), 'The Leirvík "Bønhústoftin" and the Early Christianity of the Faroe Islands, and Beyond', *Acta Islandica* 5, 27–54.

Thomsen, U.T., Drewes Nielsen, L. and Gudmundsson, H. (eds) (2005), *Social Perspectives on Mobility* (Aldershot: Ashgate).

Urry, J. (2000), *Sociology beyond Societies, Mobilities for the Twenty-first Century* (London: Routledge).

—— (2004), 'The New Mobilities Paradigm' (workshop paper).

Visjón 2015 (2007), *Infrakervið* (Tórshavn: Løgmansskrivstovan).

Chapter 7

A Travelling Fishing Village:
The Specific Conjunctions of Place

Siri Gerrard

Introduction

I have been in the fortunate situation of meeting women and men in fishing villages in Finnmark, Northern Norway since 1972, as a student, a researcher and the partner of a fisher, and gaining insights into many aspects of life there.[1] Demographic and recruitment patterns, fish produce, technology, fishing boats and fish plants have all changed during this time. People's practices thus may vary, and this indicates that places are always in the making, as Doreen Massey (1994) so nicely formulated it years ago. Kirsten Simonsen expresses the same thoughts in Chapter 2, when she writes that places are highly dynamic.

Fishing villages and their dynamic processes may be considered to be laboratories for studying places that are continuously in the making. This also implies that people's images of places are in the making or that, as expressed in the title, the images of a place resist fixity. One aspect of such a way of thinking is that people in fishing villages are on the move. Fishers move daily or seasonally. Some women and men working in fish plants, schools or other jobs leave the village to work, pursue higher education, go shopping and leave for weekends and holidays. It is this kind of mobility or travelling, and women's and men's ability to make changes, which are the focus of this article. I have been inspired by Kirsten Simonsen and her emphasis on the dynamic character of the specificity and meaning of places (see Chapter 2). I have also been inspired by a number of anthropologists who are interested in local community research, especially Trond Thuen (1997) and his discussion on how places can be 'carriers of meaning'. In this way I hope to obtain a better understanding of the dynamic and mobile character of fishing villages.

1   This chapter is a part of the project *Sustainable coastal culture? A gender perspective on resource and area use*, financed by the Research Council of Norway and its programme, *Towards Sustainable Development: Strategies, Opportunities and Challenges*. My deepest thanks go to the population of Skarsvåg, to Kirsti Pedersen Gurholt, Unnur Dís Skaptadóttir and the editors for inspiring comments. The data has been collected by means of interviews, participant observation and statistical sources.

Perspectives on the Dynamic Character of Places

It is not a simple task to reveal the dynamic character of places. In Chapter 2, Simonsen emphasizes that the constitution of place affords a continuous construction and reconstruction in all its complexity and indeterminacy, involving a conjunction of institutional, corporeal, symbolic and material elements. She also demonstrates that aspects such as embodiment, emotions, situated corporal attitudes like love, fear and mutuality, as well as narratives and memories, are elements that give rise to 'places as encounters'. Such aspects imply a focus on symbols, materialities, social practices, relations and narratives (Massey 1994; Massey 2005). In this way, places become different, unique conjunctions of networks of practices and relations, not just a set of static conditions affected by dynamic processes but co-producing constellations in continuous construction and change.

Anthropologists are also aware of the complexity of places. Thuen (1997, 277) poses the question of how places can be 'carriers of meaning' in people's cognitive orientation. Such a perspective, he says, makes it possible to look further into how a place can be made a common reference for identity and feeling of belonging through its symbolic foundation in the space of the common experience of understanding. He places emphasis on the fact that interaction may contribute to constructing and confirming the place as a point of reference for common identification by means of shared experiences, memories, meanings, knowledge, and so on. Through the communication embedded in the interaction, places contribute to confirming common views of local incidents, creating common events with a common understanding. In this way, places vary over space and time and are constructs of the sum of events that actors assign to them. In my understanding, this is another way of expressing how 'places are constituted by the way in which we interact with them', as Simonsen expresses it (Chapter 2).

Both anthropologists and geographers focus on the fact that new identifications are being constructed and deconstructed faster and faster within modern and post-modern realities (Thuen 1997, 281). Such processes are often related to globalization. With fishing villages in Finnmark as the focal point, we know that global contacts and relations are not new. Through the fishing trade, mobile fishers and commuting workers, the relation between individuals and places across local or national borders has been established for ages. Now we can ask how fishing villages are constructed and relations established and maintained in a so-called neo-liberal, borderless global economy.

I believe that such perspectives are relevant to an understanding of the complex specifics of fishery villages. People's ideas of the meaning of place have to be taken into consideration. Such ideas, for example, may last for longer or shorter periods, and may be coloured by a special event or the individuals present (Thuen 1997, 284–5). The meaning of place may also be something more. The landscape defined, for example, by means of natural resources, like fish, may have a different meaning for different people, in fishing villages as well. 'Places, like voices can be local and multiple. For each inhabitant, a place has a unique reality, one in which meaning is shared with other people and places' (Rodman 1992, 643). In this way, places create the raw material for individual or generation-specific experiences in time and space. They are, as Rodman expresses it, multi-local and may be experienced differently. Through local discourses about the interpretation of specific incidents, Thuen emphasizes that places are constituted, not as common ideas, but as a theme with variations that people consider relevant and important (Thuen 1997, 286).

In what follows, I shall focus on a number of events and see how fishery people make places into encounters or, to put it another way, how they give meaning to places. I shall use examples from Skarsvåg (see Figure 7.1), one of the northernmost fishery villages in the world, situated in the municipality of Nordkapp, Finnmark, as the locus for my arguments. I shall start by drawing on the fact that the residents have been and still are mobile. I use the concepts of mobile residents and mobility in the sense that the residents do not stay in the village all the time. They travel. Travelling as a metaphor indicates that people move between places for shorter and longer periods. Such movements may also indicate cultural and social changes. Households travel, too, with all their members journeying together. My main concern is therefore in what way a sense of place is constructed in villages with travelling individuals and households. Are there practices and institutions that counterbalance the mobility processes? Might local festivals provide an example of practices that may contribute to a sense of community? Are both travelling and community specific characteristics of fishing villages?

Figure 7.1   Skarsvåg

Source: photo by Anne Lise Lindkvist Nyheim.

Travelling Residents

Today, between 80 and 90 women, men, young people and children, living in 40
households (counted personally), have daily activities based in Skarsvåg, compared
to 220 individuals in 1970 and 1980 (SSB 1971; SSB 1981). The heavy decrease
over the past 25 years is a trend that is to be found in most coastal areas and in
municipality centres in Finnmark as well (SSB 2007; SSB 2007a). The population
increases from May to September, when the holiday residents arrive and the Sámi
women and men from Inner Finnmark settle with their grazing reindeers, and produce
and sell souvenirs from their little store in the municipality centre.

Fishing and processing have been the basis of the economy for centuries (Gerrard
1975). Today, there are about 20 active fishers harvesting cod in the winter and the
spring, coalfish and haddock in the summer and autumn. The plant has between five
and ten workers and has been owned by non-local owners since the 1930s. Now it
belongs to one of the largest companies in Norway, Aker RGI, and sends most of the
fish to the company's filleting factories.

The schoolchildren attend classes in the village from the first to the tenth grade.
They leave the village when they start high school. This is accepted, but when
the politicians speak about moving the school classes from fishing villages there
is a reaction. There are also examples of commuting mothers and children: the
mothers to work and the children to the kindergarten. The part-time day care for
children of three years of age and over, connected to the school, does not cover the
needs of the smallest children and their mothers. Elderly people receive help in
their homes until they cannot manage and have to move to the old people's home
in the municipality centre.[2] These types of activities generate jobs in the public
sector, especially for women. Adult women have educated themselves as teachers
by means of distance teacher training. Teachers and fishers attend meetings and
courses outside the village. In this way women and men, young and old, are used to
travelling professionally.

The tourist magnet, North Cape, where the tourists enjoy the view and the
midnight sun, is within 14 kilometres of the village. German men also come to fish.
New and older buildings, for example a handicraft shop 'The Christmas and Winter
House', the tourist centre in the harbour area, a hotel, a motel and campsites all
serve the tourists. However, these services do not represent sufficient possibilities
for young, unmarried women and men. They leave in order to obtain an education or
to get more stable jobs in other places. Few come back and settle.

All the shopping has to be done in the municipality centre. Once in a while,
young families also travel to the shopping centres in the nearest towns to buy special
items like party clothes and furniture. Another group of travellers is made up of those
who have partners in other places. At weekends or during the holidays, men travel
from Skarsvåg to the home they share with their female partner.

Children and adults also have to travel 27 kilometres to the municipality centre
when they need the health service or want to participate in special sports and cultural
events. However, the local branches of national associations relating to fishing,

---

2   Honningsvåg, see Chapter 5 by Paulgaard.

sport, youth activities and women's interests, the café organized by the youth club and the swimming pool and sauna in the school building represent important local meeting-places.

The mobility or travels of today are different compared with the situation 35–40 years ago, due to material, economic, political and cultural changes. Previously, roads were closed in winter and few people owned cars. Today, most adults have a driving licence so they can use the good roads that were constructed in the late 1980s, and the roads are kept open all year round. Commuting is possible. Changes in the fishery policy and the quota system, introduced in 1990, limit the number of fish that can be caught, thus creating more spare time ashore, for some at least. Women want jobs to suit their education and interests. As I mentioned earlier, it is clear that Skarsvåg today is a village where residents and others come and go. In this way the population has become used to short and long-term travel for work, education, kindergarten, shopping, leisure, health and care, as well as love.

Travelling Households

Travelling households in fishery villages in Norway are an issue that is rarely discussed or written about. This is in contrast to the Sámi reindeer herder literature (for example, Paine 1965; Ingold 1993; see Chapter 3 by Mazzullo and Ingold) and more general nomad literature (Stenning 1958). Mazzullo and Ingold argue that Sámi reindeer herders' lives are 'lived not *in* places, but *along* paths'. The dwelling space extends along 'alternately converging and diverging paths' that people take in the pursuit of their livelihoods. Places are formed on these paths where people camp or set up stores. Since the 1980s, reindeer herder households in Skarsvåg have lived in regular houses purchased from their heirs or from people who have moved away. These days, households consist mainly of middle-aged and elderly individuals who have come to the coast and stayed. The younger men come from Inner Finnmark when their labour is needed. Thus, travelling practices are undergoing generational changes. The meaning of coastal places for the travelling reindeer-herder household members seems to be influenced, amongst other things, by good grazing land, relatively easy herding and the possibility of carrying on small-scale business with the tourists.

Today, members of many households in Skarsvåg travel. Half of the households move between their domiciles in the village, where they are formally registered as inhabitants, and their own holiday homes, cabins or caravans at fixed campsites. Some of these houses have been inherited jointly with siblings. Others have bought a house or built a cabin near their relatives' homes, for example in Porsanger, the neighbouring municipality of Nordkapp, two hours' distance from Skarsvåg. During the summertime and at weekends, all the household members go to the holiday home. The older generation may stay there for longer periods, since they do not have to go to work regularly. Both women and men maintain that the climate and social relations play an important role when they choose to go away from the fishing village. Gretha told me:

There is a big difference between the climate here at the coast and in the Porsanger Fjord. In Skarsvåg it can be cold and windy. Some mornings clouds or fog cover the area. When it is 10° C here, it is almost 20° C there.[3]

Those who have holiday homes where they were born often emphasize the potential for getting together with children or relatives. Gerd gives an example:

We often visit my relatives and they come to us. This happened less often before we built our cabin. My son, who lives down south visits during the hunting season! My husband and I go skiing, pick berries and just enjoy the nature.

Linda, who is in her thirties, said that she likes to take her husband and children with her and spend time at her parents' cabin near the town where she was born, a long distance away from Skarsvåg. She talked lovingly about the mountains, the view from the cabin, she and her children meeting up with parents and grandparents. She also reflected on the fact that she might one day move back to this area. Two of the three women cited above come from an outside area. Today they consider themselves to be residents of Skarsvåg. Whether they will stay there 'forever' is difficult to say. This year, a senior household moved away for good. Women and men in the newly-established households say that they will stay until the children start high school. Some of the youngest fishers say they will stay as long as there are fish in the sea.

These examples from Skarsvåg demonstrate that relations between the domicile and the holiday home create practices and meanings that are closely related to different aspects of place, for example the former and current use of the land or seascape. In the fishing village they can obtain fresh fish and make fish cakes, which they can then take with them to the holiday home. From the holiday home or cabin they bring back berries of different kinds. What is also interesting is the emphasis on the contrasting climate of the fishing village and holiday place. It is also evident that contact with relatives living in the holiday home area is important. The holiday home thus becomes a means of sustaining relations with friends and relatives, maintaining and developing family ties (Paulgaard 1993).

Another aspect of these travelling households is the fact that they move between two well-equipped houses. In the holiday home they may not have the newest television or furniture, but most houses are equipped with domestic necessities. This is a trend that is taking place throughout the whole of Norway (Abrahamsen and Stangeland, 1996).

Another group of travelling households comprises the older generation who, since the mid-1990s, have spent between two and three months in the Mediterranean during the winter. This is something else they have in common with other parts of Norway (Kjølstad and Ovrum 2000; Helseth, Lauvlie and Sandlie 2004). The Costa del Sol in Spain, the Algarve in Portugal and Cyprus are places where many people from Finnmark go. They rent a flat and establish a household with daily routines such as shopping, cooking food, exercising and attending courses.

---

3   All translated citations are my own.

The practices on Cyprus are like those at home, but not exactly, say some of the older generation. Again, the climate is different. That is perhaps the main reason why they are there: to avoid the hardest winter months at home. They establish new friendships and new practices, for example going to courses which they would seldom have the possibility to do at home. They also say that they gain experiences to share with their children and others who travel to the Mediterranean. Some years ago, an entire household – mother, father and two children – spent a year in Alfaz del Pi in Spain (Gerrard 2003).

Women, men and entire households that have moved away come back to the village to spend their vacation in the parents' house. They renovate, visit relatives and very often go fishing. Some also go hiking in the mountains and fish in the small lakes. Households where the household members were born in the southern part of Norway also spend their holidays in Skarsvåg. Through their Skarsvåg-connected network of friends and family they have bought their houses at low prices. They talk of Skarsvåg as a place with nice, helpful neighbours. The Skarsvåg residents, for their part, refer to them as friendly people who come and renew the houses, and sometimes establish friendships. By doing so they confirm that Skarsvåg is a nice place to live.

These examples illustrate what Simonsen indicates (see Chapter 2), that fishing villages like Skarsvåg, too, are constituted by 'different, unique conjunctions of networks of practices and relations, not only a set of static conditions that the dynamic processes effect, but co-producing constellations in continuous construction and change'. The majority of the residents still relate to the facilities offered by nature, but in different ways. The relationships they build seem to vary. The number of non-local relations is increasing, compared to local relations. The residents are stable inhabitants, according to the national registration office, but they also travel. Their motives for travelling vary. What the travellers do at the places they commute between varies, as do the stories that they tell. The same may be said about the men and women of the reindeer-herder households, and of former or newly-established holiday resident households. The various groups mobilize different aspects that offer different meanings to the place. Special events such as berry-picking in the Porsanger area, or attending courses on Cyprus, stimulate special feelings and actions that may influence the meaning of place so that it varies from person to person, and household to household. Such examples add to the argument that places are dynamic and situational.

Encounters Maintaining a Sense of Community

While the examples of these travellers illustrate that fishing villages may be conceived in different but specific ways by different people in different situations, there are also events that help to create or maintain a sense of community. Festivals or 'Skarsvåg Days' arranged in 1992, 1996 and 2001 are events which may counterbalance differences in the sense or image of place which may result from individual and household travel.

The first Skarsvåg festival was dedicated to the opening of the fishery section of *Skarsvåg Fiskeri og Turistservice*, consisting of bait shacks, freezer and cold storage. The shareholders, both residents and people who had moved away, were invited by the festival committee to celebrate with a formal dinner. In 1996, they celebrated the anniversaries of Skarsvåg Chapel Association, Skarsvåg Fishermen's Association and Skarsvåg Youth and Sports Association. In 2001, the plans for new tourist institutions were presented at the formal dinner.

The festival starts on a Friday afternoon and concludes on the Sunday evening. A custom introduced during the reconstruction period after the Second World War has been reintroduced. When the owner from southern Norway came for his annual visit, he brought with him soft drinks for the children. Now the local steamer has been replaced by a fishing-boat. This event gives the opportunity for grandparents to tell this story to their grandchildren. Then the fishing competition for the children starts. One grandfather said of the competition: 'It is nice to be together with the kids and teach them to fish. You know, they live far away from here.' In 1996, the major arrangement was the celebration of the Skarsvåg Fishermen's Association's sixtieth anniversary with a banquet. Speeches, delivered by current and former representatives of the Fishermen's Association, former administrators and a headmaster who had moved away at a young age, all made by men, recalled exciting incidents. The menu was also popular. Everyday food from olden times has become today's party and banquet food. In 1992 there was *sei-mølje* (coalfish and fish-liver) and in 1996 the women served *boknafisk* (half-dried cod). The fishers had fished, hung and dried the cod. The women of *Norges Kvinne- og Familieforbund* (the Women and Family Association) were responsible for everything to do with the food and drink for two hundred guests – decorations, cooking and service, and supervising the younger family members – which was undertaken by residents and women who lived in other places alike. They served the food in a professional way. The women's knowledge of food and organizational and decoration skills are crucial to such a ceremonial arrangement.

Members of Skarsvåg Youth and Sport Association were responsible for the variety shows and the dance to the music of a dancing band from the village that followed. The people coming home on holiday were among the most enthusiastic. The dance afterwards and the conversation around the tables caused many people to remember their own youth and first dance party, but also to talk to people they had not seen for a long time.

Saturday was the day for outings. 'The whole village' clambered into the fishing boats to visit areas further out in the fjord. Fires were built, sausages grilled and coffee was served. When the outing was to Lille Skarsvåg the older people told stories they had heard from their parents and pointed out the remains of an old cemetery. When they went to Hornvika, at the foot of North Cape, they talked about the Skarsvåg families who owned the café and the children who sold flowers to the tourists. This provided a good opportunity to show children and grandchildren places for Sunday outings. On the trip to and fro, all of those who were especially interested had the opportunity to study the boats and discuss the newest plans and fishery policies. The festivals ended with a church service and coffee afterwards on the Sunday.

In Skarsvåg the festivals are encounters which women and men define, carry out and enjoy in their own way, with little outside influence. Kirsti expressed it as follows:

> Skarsvåg Days are an effort, but they are an effort that I am more than happy to be part of. Skarsvåg is still a vigorous village. The Skarsvåg Days show this. We want to celebrate this with people from the village, but also with those who have moved away. Now we are waiting for the next arrangement. Some of the people who have moved away have said for many years that it is their turn to take the lead.

Jørgen, a fisher, expressed himself as follows:

> The festivals are important because we have to collaborate. We do things together and give attention to our village. It is a period with a lot of hustle and bustle. Some of former residents come home. What does it mean in the long term? Well, some have bought their holiday home here.

Hope for the future is evident in Skarsvåg. This represents a contrast to Newfoundland, where emphasis has been placed on the home-comers and celebrating the 'good old times' (Powers 1998).

The significance of creating something, doing something together, for others and for oneself, is manifested by a large mobilization on the part of the festival committee. Participants are of all ages and many have a specific task to perform. There is a tacit requirement for everybody to participate. However, nobody questions the fact that the Sámi women and men seldom participate. One of the young men I talked to said that he liked to go to the dance only. This is in contrast to many of the adult residents, who look upon the festival as a sign of development and viability, despite the decline in population (Gerrard 2000). The children have the opportunity to be with their relatives, to do lots of activities with adults and gain some insight into the history and old memories. For the people who have moved, the festivals represent their roots and a place to spend holidays or 'holiday country' (*ferielandet*) (Paulgaard 1993). Since 2003, Skarsvåg women living in the eastern part of Norway have decided that they will not wait for the next festival. Now they meet annually, not in Skarsvåg, but closer to where they live.

Even though the last festival was arranged some years ago, people are still talking about them as important events and looking forward to a next one. In this way, the festival practices are manifested in narratives or stories. The statements above demonstrate that they are important to both women and men: being together and collaboration are highlighted. Relations towards the people who have left also seem to be important. This was not the case during the 1950s and 1960s. At that time, long-distance travel took time and was expensive. At that time, I used to hear fishers say that the people who left – regardless of what they did in their professional life – were not as important as those who stayed (Gerrard 2003). I therefore maintain that the conjunction of networks of practices, relations and the value placed on events such as festivals both contribute to what Massey (1995) has labelled co-producing constellations in continuous construction and change. It demonstrates how the festivals have become a symbol of a vivid and strong fishery village for the residents, as well as 'the good old days' for the holiday residents. My understanding is that this is another expression of places as encounters, as defined by Simonsen in Chapter 2.

The Specific Characteristics of a Fishery Village: Mobility and Community?

With an emphasis on individual and household travel, one might expect residents to consider Skarsvåg's sense of place in terms of mobility. As demonstrated above, this is not necessarily the case. Many of the residents still think of Skarsvåg as an active village where, throughout the year, they have succeeded in obtaining a local health service, a better harbour and quays, a school with a swimming pool, better roads, and now also better fishery and tourist facilities (Gerrard 1975; Gerrard 2000; Gerrard 2003). The history and the fact that there are still fishery activities going on, even more so than in the municipality centre, and the festivals and stories about them, seem to counterbalance the mobility tendency and maintain Skarsvåg's sense of place as a community. There are still associations, unions and local projects, for example at the school, although the activities and the number of members vary. Women who have moved away also meet up.

In daily life, people seem to worry less about short-term and short-distance travel, or what may also be termed everyday mobility. They worry about a heavy decline in the population and the threat of losing school classes. Travel to and fro, everyday mobility in established and new forms, is like doxa: it is taken for granted and seldom problematized. Even long-term mobility, including the people who have moved away, is not thought of as 'final' in the way it was before. Children and relatives keep in touch by telephone and the Internet (see Chapter 8 by Larsen and Urry). By participating in festivals and expressing their admiration for local food and local nature they all confirm their village identity. The festivals and other meetings of a collective nature contribute to making the place into an encounter (cf. Chapter 2). The people who have left retain elements of Skarsvåg in their memories and topics for small-talk, like the women who meet up once a year. Once in a while, however, when festivals are arranged in Skarsvåg, it is important for both residents and those who have moved to attend with their families. In this way it is evident that once in a while, but not all the time, corporal placement in the village becomes important.

The interaction between residents and others may be viewed as a contribution to constructing and confirming the place as a point of reference for common identification by means of shared experiences, memories, meanings and knowledge, as Thuen (1997) has expressed it. Collective interaction and the exchange of opinions may lead to common views, for example about the importance of festivals. I therefore maintain that the fishing village as an encounter becomes a combination of realities and stories about past events, present-day actions and hopes for the future.

It seems to me that Skarsvåg, as a fishing village, is a place with varying material realities, but there have been efforts to create encounters, like the festivals. In this way, the meaning can be shared in a wider context, and with people who today live in other places. The fishery village, with its structure and culture, creates the raw material for individual or generation-specific experiences in time and space. The mobility of the individuals and the households strengthens the sense of place as multilocal and multivocal, to use Rodman's (1992) terms. In addition, fishery villages seem to be constituted not as common ideas but as a theme with variations that women, men and children residing in, or with roots in, the village consider to be relevant and important (Thuen 1997, 286). Seen from this angle, fishing villages, too, turn out to be a concept of difference.

I have argued that the mobility aspects represented by travelling individuals and households can be counterbalanced by the community aspects represented by festivals, talking about them, meetings, conversations on cell phones and on the Internet. These stories function like glue, binding events and people together. This is also demonstrated by the women's meetings in the southern part of Norway. The experiences and the identity they have developed in relation to their place of birth, as well as the exchange of memories, opinions and knowledge, are just as important as the meetings in Skarsvåg. When they meet they confirm this, and even construct new meanings. For them, Skarsvåg is not necessarily the most important actual meeting-place.

Several researchers have discussed this type of construction. Benedict Anderson (1983) has written about imagined communities. Stuart Hall (1995, 182) emphasizes that communities include the idea we have of them, the images we use to conceptualize them, the meaning we associate with them, and the sense of community with others that we carry inside us. These are the ways in which communities are imagined – and the way in which we give meaning to the idea of community. This could also be said to be the case in Skarsvåg. It seems to me that the stories about the festivals and other meetings have become more and more important, the more mobile the residents have become.

References

Abrahamsen, H. and Stangeland, G. (1996), *På hytta – vårt andre hjem* (Oslo: Boksenteret).

Andersen, B. (1983), *Imagined Communities: Reflections on the Origins and Spread of Nationalism* (London: Verso).

Fossåskaret, E., Fuglestad, O.L. and Aase, T.H. (eds) (1997), *Metodisk feltarbeid. Produksjon og tolkning av kvalitative data* (Oslo: Universitetsforlaget).

Gerrard, S. (1975), *Arbeidsliv og lokalsamfunn: Samarbeid og skille mellom yrkesgrupper i et nord-norsk fiskevær* (Tromsø: Institutt for samfunnsvitenskap, Universitetet i Tromsø, mag. art. dissertation).

—— (2000), 'The Gender Dimension of Local Festivals: The Fishery Crisis and Women's and Men's Political Actions in North Norwegian Communities', *Women's Study International Forum* 23:3, 299–309.

—— (2003), 'Må det bo folk i husan. Nye levekår, nye levemåter og nye utfordringer', in Haugen and Stræte (eds).

Goody, J. (ed.) (1958), *The Developmental Cycle of Domestic Groups* (Cambridge: Cambridge University Press).

Hall, S. (1995), 'New Cultures for Old?' in Massey and Jess (eds).

Haugen, M.S. and Stræte, E.P. (eds) (2003), *Ut i verden og inn i bygda: Festskrift til Reidar Almås* (Trondheim: Tapir akademiske forlag).

Heggen, K., Myklebust, J.O. and Øya, T. (eds) (1993), *Ungdom i lokalmiljø. Perspektiv frå pedagogikk, sosiologi, antropologi og demografi* (Oslo: Det Norske Samlaget).

Helseth, A., Lauvlie, M. and Sandlie, H.C. (2004), *Norske pensjonister og kommuner i Spania* (Oslo: Nova rapportserie 2004: 3).

Ingold, T. (1993), 'The Temporality of the Landscape', *World Archaeology* 25:2, 152–74.

Kjølstad, S. and Ovrum, H.H. (2000), 'Norske turisters bruk av medisinske tjenester i utlandet', *Tidsskrift for den norske lægeforening* 2000, 120, 1991–1994.

Massey, D. (1994), *Space, Place and Gender* (Oxford: Polity).

—— (2005), *For Space* (London: Sage).

Massey, D. and Jess, P. (eds) (1995), *A Place in the World? Places, Cultures and Globalization* (Oxford: Oxford University Press in association with the Open University).

Paine, R. (1965), *Coast Lapp Society* (Tromsø: Tromsø Museum).

Paulgaard, G. (1993), 'Nye tider – nye toner? Eit fiskerisamfunn i endring', in Heggen, Myklebust and Øya (eds).

Pedersen, K. (1999), *'Det har bare vært naturlig': Friluftsliv, kjønn og kulturelle brytninger* (Oslo: Norges Idrettshøyskole, PhD. thesis).

Powers, A.M. (1998), 'Come home year celebrations as pilgrimage', Paper presented to the British Association of Community Studies, Stoke on Trent, England.

Rodman, M.C. (1992), 'Empowering Place: Multilocality and Multivocality', *American Anthropologist* 94:3, 640–56.

Statistisk sentralbyrå (SSB) (1971), *Folke- og boligtellingen. Kommunehefte for Nordkapp* (Oslo: SSB).

—— (SSB) (1981), *Folke- og boligtellingen. Kommunehefte for Nordkapp* (Oslo: SSB).

—— (SSB) (2007), *Folke- og boligtellingen 2001, Kommunehefte for Nordkapp*, http://www.ssb.no/emner/02/01/10/beftett/.

—— (SSB) (2007a), *Folkemengde etter fylke og kommune*, http://www.ssb.no/fob/emner/02/03/folkefram/tab-2005-12-15-09.html.

Stenning, Derrick J. (1958), 'Household Viability Among the Pastoral Fulani', in Goody (ed.).

Thuen, T. (1997), 'To perspektiver på studiet av lokal samfunn', in Fossåskaret, Fuglestad and Aase (eds).

PART 2
Connections and Encounters

Chapter 8

Networking in Mobile Societies[1]

Jonas Larsen and John Urry

Introduction

Modern societies are paradoxically typified by 'time-space *compression*' (Harvey 1989) and 'time-space *distanciation*' (Giddens 1990). Distances have never meant so little *and* so much. The world seems to get smaller *and* larger at the very same time. It is getting compressed or smaller because innovations in transport and communications constantly reduce the time (and money) it takes to overcome physical distance. Email is a striking recent example of 'time-space compression': travelling great distances within nanoseconds.

But new technologies also increase distance. 'Time-space compression' can involve more spatially dispersed social networks, as close ties increasingly live in 'distant' places. Recent studies show a shift from 'little boxes' of spatially dense and overlapping social ties to spatially dispersed *networks* (Wellman 2002; Larsen, Urry and Axhausen 2006). Such a shift stems from historically high levels of movement for jobs, education, holidays and international migration, as well as of communications making it possible *and* necessary to network at-a-distance. Even those who are 'immobile' deal with distance if a friend or family member moves to a far-off place. 'Time-space distanciation' captures how remote interaction has become an increasingly significant feature of social life. When networks are distanciated it is harder to meet up, to network *physically*. While they can connect instantly and effortlessly by phone or email, one's 'strong ties' often spend hours or days travelling in cars, trains and planes just to spend brief moments of 'quality time', and such travel is not cost-free. Even though travel times and costs have shrunk within the last decade, Kellerman reminds us that 'international travel still implies the existence of a friction of distance, expressed in time, money, effort, whereas the movement of information is almost free of such frictions' (2006, 85–6).

This chapter builds upon the assumption that social networks only *work* if they are performed regularly in and through texting, phone calls, emails, blogs, home pages and so on. Without such networking at-a-distance, the distances that separate people can be unbearable. This chapter discusses how modern networking involves complex combinations of face-to-face and face-to-interface interaction, at-near and at-a-distance.

---

1 The first part of this chapter in part draws upon *Mobilities, Networks, Geographies* (Larsen, Urry and Axhausen 2006) and *Mobilities* (Urry 2007).

Most social science has not perceived physical distance as a problem or even as particularly interesting (except for transport studies/geography) (but see Bærenholdt 2007). We treat physical distance, and the complex mix of presence and absence that this entails, as hugely significant for contemporary social life. This chapter explores how distanciated network members are *forced* to deal with the physical distances that separate them (on how people cope with *cultural* distances, see Bærenholdt 2007). We analyse how, and by what technological means, networking takes place in mobile network societies and (re)produces networks. The chapter takes its inspiration from actor-network theory, by viewing contemporary networking practices as heterogeneous, and various material technologies and virtual places form part of that heterogeneity. Social networks are intricately networked with extensive material networks (Licoppe and Smoreda 2005). Overall, we argue for a *practice* approach that understands social networks as mobile and performed, having to be practised in order to be meaningful and durable. Networks should be viewed as an accomplishment, involving and made possible by various 'network tools' (see Larsen, Urry and Axhausen 2006). The chapter deploys the concept of *network capital*, by discussing empirical research of networking practices at-a-distance among poorer transnational migrant families.

We begin by outlining briefly why it is appropriate to speak of contemporary social life as *networked*.

Networked Life

Castells argues that contemporary societies are 'network societies'. This is a society made up of networks powered by micro-electronics-based information and communication technologies:

> What is specific to our world is the extension and augmentation of the body and mind of human subjects in networks of interaction powered by micro-electronics-based, software-operated, communication technologies. These technologies are increasingly diffused throughout the entire realm of human activity by growing miniaturization [and portability]. (2004, 7)

In particular, 'networked computers', 'mobile telephony' and fast transportation have been crucial in producing this global network society of 'timeless time' and 'spaces of flows'. The latter highlights how dealing with distances and networking at a distance have become crucial:

> Simply put, the *space of flows* is the material organization of simultaneous social interaction at a distance by networking communication, with the technical support of telecommunications, interactive communication systems, and fast transportations. (Castells et al. 2007, 171)

Castells' account is a macro one, and he does not address the network society in relation to friendship and family life for *social* networks. Yet this notion suggests that 'sociality' is increasingly networked with mobile communication, and performed through interfaces and phonescapes, rather than purely face-to-face.

Western households are increasingly networked. Families are plugged into an ever-expanding array of communications that connect family members to one another and to the outside world. These include postal systems, radio, TV, satellite TV, land-line phones and more recently mobile phones, computers, digital cameras, email accounts, home pages and blogs. For instance, 85 per cent of Danish households have one or more computers and 80 per cent are connected to the Internet.[2] The home has become a communication and information hub (Wellman et al. 2005).

In addition, 'individuals' are themselves 'networked'; this is particularly evident with regard to mobile phones affording 'person-to-person' connectivity or 'networked individualism', which 'suits and reinforces mobile lifestyles and physically dispersed relationships' (Wellman 2001, 239). 'The person has become the portal' (Wellman 2001, 238). Each person is the engineer of his/her own ties and networks, and more or less always connected (batteries and masts permitting), even on the move. As Licoppe reports, 'the mobile phone is portable, to the extent of seeming to be an extension of its owner, a personal object constantly there, at hand…Wherever they go, individuals seem to carry their network of connections which could be activated telephonically at any moment' (2004, 139). Mobiles are almost prosthetic, physically conterminous with one's body, and connect with one's networks. Widespread mobile phone ownership enables individualized yet connected small worlds of communication, in the midst of absence, distance and disconnection. Phone calls on the move and impromptu text messages have become crucial networking practices in many countries around the world, especially in Europe. In 2004, nine out of ten people in EU countries (and Norway) were mobile phone users (Castells et al. 2007). Here, virtually all teenagers and young adults own a mobile phone (Ling 2004, 16).[3]

Many individuals are networked with virtual sites, they network virtually and their 'profile' is available online. Many families now have a 'blog' or home page where they tell their news and exhibit their photographs and accounts of themselves.

The last couple of years have also seen an astonishing increase in networking on the internet – measured in terms of users of social networking sites.[4] As their names reveal, these are primarily about networking with old and new friends, and such sites are primarily – but not exclusively – for teenagers and youngish people. Some of the major ones are: *Bebo* (22 million); *Classmates* (40 million); *Friends Reunited* (12 million); *Friendster* (29 million); *Reunion* (25 million) and not least *MySpace*. Despite being only a few years old, *MySpace* now has an astonishing 'population' of at least 155 million and it is said to increase each day by around 300,000.[5] It describes itself as a 'place for friends', a place made up of user-submitted personal profiles, blogs, groups, photos, music and videos. At the core are profiles allowing people to present themselves photographically and express interests and tastes in

---

2 http://www.dst.dk/Statistik/Nyt/Emneopdelt.aspx?si=5&msi=6.

3 However, it should be noted that only one third of all countries have a penetration rate of mobiles over 30 per cent.

4 See lists of numbers of users of 'notable social networking sites' in *Wikipedia*: http://en.wikipedia.org/wiki/List_of_social_networking_websites.

5 http://en.wikipedia.org/wiki/Myspace.

music, literature, fashion and so on. Active networking takes place by connecting profiles and people can traverse the network through them (cf. myspace.com).

To be networked with such communications and virtual sites means being connected to, or at home with, 'sites' across the world – while simultaneously such sites can monitor, observe and trace each inhabited machine. As Zook et al. say, 'There is a noticeable "Google effect" as information can be tracked down much more effectively, fast dissolving the accepted notion of "privacy through obscurity"' (2004, 172). Such mobile machines thus reconfigure humans as physically moving bodies and as bits of mobile information and image, as individuals existing both through, and beyond, their mobile bodies, as emails, text messages, 'blog profiles', family home pages and so on. These inhabiting machines enable people to be more readily mobile through space, or to stay in one place, because of the capacity for 'self-retrieval' of personal information at other times or in other spaces.

Such 'networked machines' transform the sense of who is near or far, present or absent, affording networking at-a-distance. As Callon and Law maintain more generally, 'presence is not reducible to co-presence...co-presence is both a location and a relation' (2004, 6, 9). Mobile phone cultures generate small worlds of perpetual catching-up and small-talk on the move, blurring distinctions between presence and absence. Research suggests that mobile phoning and texting are about networked gossiping, 'anytime, anyplace, anywhere', of living in 'connected presence' with one's more or less dispersed social networks (Licoppe 2004; Fox 2001). Perpetual gossip at-a-distance helps people to come to terms with living in distanciated networks, where people physically bump into each other less often. Even when people are absent they can remain in communicative propinquity to their social networks.

The social sciences – especially sociology – have been overly-focused on ongoing geographically propinquitous communities, based on more or less face-to-face social interactions with those who are co-present. They have largely equated *near* with *close* and *meaningful*, while mobility has been associated with a lack of connection and commitment. Much social science has been typified by 'sedentarist metaphysics' and 'metaphysics of presence' (Creswell 2006; Urry 2007).

But we have seen how in network societies, many connections and much networking are not based upon physical propinquity of place. There are multiple forms of 'communicative presence' that occur through communications, images, messages and interactive sites such as blogs, home pages and social networking sites. Now that it is common to have significant others at-a-distance, such networking technologies' affordances in forging virtual propinquity are crucial to sustaining close relationships. Moreover, social life involves continual processes of shifting between being present with others (at work, home, leisure and so on) and being distant. And yet when there is absence there may be 'imagined presence', depending upon the multiple connections between peoples and places.

While mediated sociality is of increasing significance to sustaining and forging network ties, networking practices are rarely purely mediated. Face-to-face interaction is still valued and can occasionally be accomplished. Following the work of Wellman and his collaborators, highlighting how Internet communication amplifies rather than replaces face-to-face interaction (Wellman 2001, 242; Hampton and Wellman 2001), we may say that networking practices 'are complex dances of face-to-face encounters, scheduled get-togethers, dyadic telephone calls, emails to one person or several, and broader online discussions among those sharing interests' (Wellman 2001, 11).

In the light of this, Castells' account of the 'network society' appears overly cognitivist, since it suggests a society of foot-loose 'informationalism' that bypasses places, face-to-face sociality and human net*working*. Yet networks are about net*working*, and such networking involves the embodied work of travelling and meeting face-to-face to talk, work and enjoy the company of 'others'. Networking contingently produces networks (see Bærenholdt 2007). A network only functions if it is intermittently 'activated' through occasioned co-presence. All other things being equal, 'network activation' occurs if there are periodic events each week, or month, or year, when meetingness is more or less obligatory. 'Co-present interaction' is fundamental to social interaction within companies, families and friendships, for producing trust, sustaining intimacy, having pleasurable gatherings and meeting obligations (Boden 1994). These are often unmissable, obligatory and make or *perform* the network in question. In order to continue to exist within a given network there are obligations to travel, to meet up and to converse, as we see below.

Network Capital

We have argued that networking and social relations depend more and more on communications and occasional travel. This also means that 'network capital' – comprising *access* to communication technologies, transport, meeting places *and* the social and technical *skills* of networking – is crucial to the character of modern societies. Network capital is the capacity to engender and sustain social relations with individuals who are not necessarily proximate, which generates emotional, financial and practical benefit. 'Network capital' refers to a person's, or group's, or society's *facility* for 'self-directed' corporeal movement and communication at-a-distance.

We call it network capital in order to bring out the fact that the underlying mobilities do nothing in themselves. Network capital therefore points to the real and potential social relations that mobilities afford. It is the 'social relations of circulation' that are key. Network capital thus *cannot* be separated from embodied network practices. Cars, mobile phones, computers and email accounts are increasingly *necessary* for social life, but we have stressed how it is the 'hybrid performances' of talking, driving, phoning, texting, emailing and blogging that create networks. Instead of focusing on the formal structures of networks, or affordances of mobilities, we need to examine the embodied, hybridized construction of networks and practices of networking. Social networks come to life and are sustained by various practices of networking through email, forwarding messages, texting, sharing gossip, performing meetings, making two-minute bumping-into-people conversation, attending conferences, cruising at receptions, chatting over a coffee, meeting up for a drink and spending many hours on trains, or on the road, or in the air in order to meet up with business partners and clients, as well as displaced friends, family members, workmates and partners. Networking is effectively work: sometimes tedious and tiring, sometimes enjoyable and stimulating. This approach comprehends social networks as something accomplished, in process, weaving together the material and the social, as well as pleasures, obligations and burdens. Travel, meetings, writing and talking make networks come intermittently to life.

Network capital comprises eight interdependent elements:

1. *An array of appropriate documents, visas, money, qualifications*, which enable safe movement of one's body from one place, city or country to another.
2. *Others (workmates, friends, family members) at-a-distance*, who offer invitations, hospitality and meetings, so that places and networks can be maintained through intermittent visits and communications.
3. *Movement and communication competences*: walking, cycling or driving distances between different environments; to be able to see and to board different means of mobility; to be able to carry or move baggage; to read timetabled information and speak foreign languages (especially English); to be able to access computerized information; to arrange and re-arrange connections and meetings; the ability to and competence and interest in using mobile phones, text/photo messaging, email, blogs, home pages, Skype, and so on.
4. *Location-free information and contact points*: fixed or moving sites where information and communications can arrive and be stored and retrieved, including real/electronic diaries, address book, answerphone, secretary, office, answering service, email, web sites, mobile phones.
5. *Communication devices*: to make and remake arrangements, especially on the move and in conjunction with others who may also be on the move.
6. *Appropriate, safe and secure meeting/networking places*: both en route and at the destination(s), including office, club space, hotel, home, public spaces, street corner, (internet) café, interspaces, ensuring that the body is not exposed to physical or emotional violence.
7. *Physical access* to cars, road space, fuel, lifts, aircraft, trains, ships, taxis, buses, trams, minibuses, email account, internet, telephone, and so on.
8. *Time and other resources to manage and co-ordinate points 1–7*, especially when there is a system failure, as will intermittently happen.

Thus, 'network capital' complexly comprises technical, cognitive and social skills and depends on 'access' to technical, cultural, social, economic and environmental resources. 'Network capital' is a *relational* possession that depends on other people's network capital. For instance, mobile phones and email accounts are of little worth if one's ties lack or refuse these possessions. So network capital is not an attribute of individual subjects but it is a product of the relationality of people with other people and the affordances of the 'environment'. People are part of networks that both enable and constrain possible actions. They are immobilized and mobilized in complex relational ways.

We shall now discuss our notion of network capital with relation to Putnam, who has developed a related concept of social capital that 'refers to connections among individuals – social networks and norms of reciprocity and trustworthiness that arise from them' (Putnam 2000, 19; Urry 2002; Larsen, Urry and Axhausen 2006,

Chapter 2). Putnam perceives such capital as being fostered within propinquitous communities. Such communities, with high social capital, are characterized by dense networks of reciprocal social relations; well-developed sets of mutual obligations; generalized reciprocity; high levels of trust in their neighbours; overlapping conversational grouping; and bonds that bridge conventional social divides. Social bonds, and especially involvement in civic work within neighbourhoods, generate social capital.

From our point of view, the main problem with Putnam's account is that it presumes that only small-scale, localized communities can generate social capital and it fails to understand the significance of travel and networking at-a-distance to the production of social capital in mobile network societies. Putnam outlines how to reverse declining local social capital. One suggestion is this: 'Let us act to ensure that by 2010 Americans will spend less time traveling and more time connecting with our neighbors than we do today...' (Putnam 2000, 407–8).

Yet it seems implausible to argue today that trust and reciprocity are only generated within propinquitous communities. As Mason argues, 'geographical proximity or distance do not correlate straightforwardly with how emotionally close relatives feel to one another, nor indeed how far relatives will provide support or care for each other' (Mason 2004, 421). Kinship and migration researchers have shown that presence and absence – or proximity and distance – do not necessarily conflict. Indeed, intimacy and caring take place at-a-distance, through letters, packets, photographs, emails, money transactions, telephone calls and recurrent visits. Social capital stretches over long geographical distances if there is appropriate network capital. Thus, intimate and caring networking does not necessarily imply co-presence or face-to-face proximity: people can be near, in touch and together, even when great distances tear them physically apart. Network capital enables disconnected people to connect, to produce social capital. This capital brings out the way in which co-presence and trust can be generated at-a-distance and it presupposes extensive and predictable travel and communications.

As people are distributed 'far and wide', so travel and networking at-a-distance are essential to family and social life. People have to talk and come together from time to time; this is obligatory and costly. Networking is crucial and there are huge inequalities of access to the resources, the network capital, through which various forms of networking are performed and social capital is realized. Those networks high in network capital enjoy significant advantages within the systems of social inequality operating in the contemporary world. Those without sufficient network capital will have problems sustaining close ties at a distance and their social capital will suffer as a consequence. Network capital – like mobility – is caught in various power geometries of everyday life, within and across countries (Massey 1994; Cresswell 2006; Hannam, Sheller and Urry 2006), as we shall now show, using one empirical example.

In the next section we shall discuss 'anthropological' mobility research that shows how 'poor' transnational migrant families are *forced* to network at-a-distance, sometimes with very little 'network capital'.

Transnational Migrant Networking

The number of international migrants doubled worldwide between 1960 and 2000 (UNDP 2004, 87). Such migration is rarely an isolated decision pursued by individual agents, but a collective action involving families, kin and other communal contacts. Migrants travel to join established groups of settlers, who provide transnational arrangements for them in receiving countries, while simultaneously retaining links with their country of origin and with chains of other migrants (Goulborne 1999; Salaff, Fong and Siu-Lin 1999). Migration literature also examines those left behind: family members who have to learn to live with their 'strong ties' at-a-distance and networking across borders through forms of communication. Migration is a far from one-way journey involving leaving one's homeland behind: it is often a two-way 'journey' or circulation between two sets of 'homes' (Ahmed et al. 2003), with 'national' network members in various other countries (see Larsen, Urry and Axhausen 2006, 87–9). The migration experience is also one of intensive networking at-a-distance, often on a daily basis:

> *Regular* communication – whether through telephone calls, remittances, letters, voice recordings, SMS messages, photographs or visits – is part and parcel of *everyday* life in transnational families... (Parreñas 2005, 317; our emphases)

Without neglecting the unequal distribution of network capital globally, it is important to note how migrants are often some of the first to adopt new technologies because their networking relies upon their affordances (Horst 2006).

One such recent technology is the Internet – now with between 700 million and 1 billion users worldwide. Thus in Trinidad, where about 60 per cent of families contain at least one migrated family member, the Internet is the crucial networking tool 'that has permeated all sectors of society, from yachties to squatters' (Miller and Slater 2000, 12, 36). Hiller and Franz discuss how diasporic web sites and blogs are important 'meeting places' in gaining 'new ties', sustaining 'old ties' and recuperating 'lost ties', both with people in the homeland and others in the diaspora (2004).

While the Internet is crucial for many 'richer' migrants living in Europe and North America, telephone calls are of greater significance to non-elite migrants (Vertovec 2004, 219). Many 'poorer' migrants do not have access to the Internet at home or at work, and even if they do, their friends and family members 'back home' often lack it (thus network capital is relational).

The overall volume of international telephone calls increased at least tenfold between 1982 and 2001 (Vertovec 2004, 219). This happened in most countries, partly through innovations such as pre-paid phonecards and mobile phones. Thus, for migrants and their kin within the world today, 'transnational connectivity through cheap telephone calls is at the heart of their lives' (Vertovec 2004, 223). Pre-paid phonecards and mobile phones have increased the regularity and length of phone calls made by transnational migrant families across borders; in addition to this, transnational text messaging has become a mundane, *everyday* practice. According to *Wikipedia*, by 2005 there were more than 2.1 billion mobile phone subscribers (compared with 1,263 million landlines), with Africa having the fastest growth rate.

In Jamaica, 86 per cent of those over the age of 15 own a mobile phone and these are largely used for long-distance and international calls by both the wealthy and the poor (Horst 2006, 144). The 'innovation' of pre-paid phonecards has made the cost of internal calls much cheaper and more manageable, and the subscriber does not have to commit to a long term contract; this is one major reason why transnational calls and 'texts' by (mobile) phone have increased tremendously amongst migrants and their families. In general, pre-paid phonecards have 'allowed those without credit history, a permanent address, or a stable source of income to purchase cell phones' (Uy-Tioco, cited in Castells et al. 2007, 20). In addition, the 'mobile phone system' has made telephony more widespread and accessible, especially in rural areas where landline phones were a scarce resource. People often used to have to make private calls at friends' or relatives' houses, or use busy public phoneboxes: 'as a result, phones were rare, brief and sometimes cryptic in an attempt to avoid the risk of local gossip about personal or family business' (Horst 2006, 143). By 2006, 80 per cent of the world's population had mobile phone coverage.

Pre-paid phonecards and mobile phones have thus made it cheaper and easier to engage in intimate and meaningful longer conversations *and* to make short impromptu or emergency calls to ensure 'connected presence'. We may thus speak of different types of mobile phone communications; ethnographic research shows both are central to transnational family life.

Parreñas (2005) shows how mobile phones are crucial to the way in which 'absent' Filipino mothers 'mother' at-a-distance. Mainly due to 'pre-paid mobile phone systems', 33 million Filipinos were mobile phone subscribers in 2004 (Castells et al. 2007, 22). More than a quarter of all children in the Philippines grow up with at least one of their parents living abroad. Migrant mothers do not desert their children or pass them on to relatives. Instead they are 'there' through routinized communication, which may include almost *daily* text messages:

> Migrant mothers…rely on sending an SMS to communicate with their children on a daily basis. Some children even told me that they wake up to biblical messages from their migrant mothers every morning…Sending text messages is one system mothers use to make sure that their children are ready for school in the mornings. (Parreñas 2005, 328)

In addition to text messages and short phone calls, such transnational family lives are characterized by and enacted through *longer* telephone *conversations*, normally taking place once a week, at a particular time and day (especially Sundays) (see Wilding 2006). Such lengthy turn-taking telephone meetings, where news is exchanged, troubles are talked through and solidarity is expressed, are crucial in bonding migrant mothers and their children who are at home.

While these studies show how new communications have afforded cheaper, easier and more intimate transnational family life, they also highlight the way in which such families suffer from a lack of physical co-presence: 'The joys of physical contact, the emotional security of physical presence, and the familiarity allowed by physical proximity are denied to many transnational family members' (Parreñas 2005, 333).

And yet other research shows that tourist travel to visit relatives and sustain national identity is central to the transnational life of 'richer' immigrants. Ali and Holden outline the institutions or 'system' of such 'return visits'. Across Europe and North America, diasporic cultures have their own travel agents, tour operators and international airlines (2006, 233). This is because 'many forms of migration', as Williams and Hall say, 'generate tourism flows, in particular through the geographical extension of friendship, and kinship networks may become poles of tourist flows, while they themselves become tourists in returning to visit friends and relations in their areas of origin' (2000, 7; see Coles and Timothy 2004; Williams et al. 2000, 40–41). In Trinidad it is said that one can really only be a proper 'Trini' by going abroad and occasionally returning home to visit friends and kin since, as mentioned earlier, about 60 per cent of nuclear families are thought to have at least one family member living abroad (Miller and Slater 2000, 12, 36). Various studies show how many immigrants and their (grand) children regularly visit their 'homeland' and other displaced family members across the world to keep their 'national' belonging and family networks 'alive' (Mason, 2004; Sutton, 2004). Duval (2004a, 2004b) and Ali and Holden (2006) illustrate how parents of Caribbean and Pakistani origin feel obliged to travel to their homeland and personally introduce their children to its key features. Social obligations to travel are often intricately intertwined with obligations to visit specific monuments and religious sites. Ali and Holden call this 'the myth of return' (2006).

Conclusion

This chapter thus shows that there is *not* a death of distance in network societies, because social networks depend upon face-to-face meetings and such meetings presuppose 'costly' and time-consuming travel. Increasing distances between network members would matter little if networking could be performed only on the screen and by phone, but this is not always the case. Distances matter because there are barriers to necessary, obligatory and desirable face-to-face networking. If people live a long way from their 'close ties' they are likely to see them less often than they would ideally like. This can mean a lack of sociality and general support (for example with babysitting), and it can be costly and difficult to attend 'obligatory' birthdays, weddings, Christmas and New Year parties and so on. In 'rich' societies, telephone calls, text messages or courier-delivered flowers are only substitutes for a journey to, and physical presence at, a church, hospital or Christmas dinner if people have a (very) good excuse for not being able to attend. Communications will often be too one-dimensional to fulfil significant social obligations. As Wilding says:

> Although 'connected presence' gives the appearance of the annihilation of distance, it can also result in increased guilt and anxiety when the distance becomes evident again through tragedy…In some circumstances, a telephone call or email is simply not sufficient to show care for kin in need. In some respects, the connections enabled by email and other ICTs are 'sunny day' technologies. (2006, 134)

We have also seen how amongst poorer families coming from the 'south' there is a fair amount of 'substitution' of physical co-presence, especially with daily phone calls and texting, combined with more infrequent longer calls.

Overall, we have seen just how significant network capital is in structuring the character of contemporary social life. As Bauman argues, 'Speed of movement has today become a major, perhaps the paramount, factor of social stratification and the hierarchy of domination' (2000, 151), where this speed refers not only to physical but also to communicative and virtual movement across awesome distances. We argued that social capital – or connections between people – increasingly depends on network capital, since many social networks are far-flung. Elsewhere, we have examined just how future climate change, or 'global warming', will further transform the structuring of network capital and the lineaments of a networked life, where network capital may be even more significant in producing a 'good life' only for some (Lovelock 2006; Urry 2007).

References

Ahmed, S., Castañeda, C., Fortier, A. and Sheller, M. (eds) (2003), *Uprootings/ Regroundings: Questions of Home and Migration* (Oxford: Berg).

Ali, N. and Holden, A. (2006), 'Post-colonial Pakistani Mobilities: the Embodiment of the Myth of Return in Tourism', *Mobilities* 1:2, 217–42.

Bauman, Z. (2000), *Liquid Modernity* (Cambridge: Polity).

Bærenholdt, J.O. (2007), *Coping with Distances: Producing Nordic Atlantic Societies* (Oxford: Berghahn).

Boden, D. (1994), *The Business of Talk: Organisations in Action* (Cambridge: Polity).

Bourdieu, P. (1984), *Distinction. A Social Critique of the Judgment of Taste* (London: Routledge and Kegan Paul).

Cairncross, F. (1997), *The Death of Distance* (London: Orion Business Books).

Callon, M. and Law, J. (2004), 'Guest Ediorial', *Environment and Planning D: Society and Space*, 22:3–11.

Castells, M. (2004), 'Informationalism, Networks, and the Network Society: a Theoretical Blueprint', in Castells (ed.).

—— (ed.) (2004), *The Network Society: a Cross-cultural Perspective* (Cheltenham: Edward Elgar).

Castells, M., Fernádes-Ardèvol, M., Qiu, J. and Sey, A. (2007), *Mobile Communication and Society: A Global Perspective* (Massachusetts: MIT).

Coles, T. and Timothy, D.J. (2004), *Tourism, Diasporas and Space* (London: Routledge).

Creswell, T. (2006), *On the Move* (London: Routledge).

Duval, T.Y. (2004a), 'Linking Return Visits and Return Migration among Commonwealth Eastern Caribbean Migrants in Toronto', *Global networks* 4: 51–8.

—— (2004b), 'Conceptualilsing Return Visits: A Transnational Perspective', in Coles and Timothy (eds).

Fox, K. (2001), *Evolution, Alienation and Gossip: The Role of Mobile Telecommunications in the 21st Century* (Oxford: Social Issues Research Centre).

Giddens, A. (1990), *The Consequences of Modernity* (Stanford: Stanford University Press).

Goulborne, H. (1999), 'The Transnational Character of Caribbean Kinship in Britain', in S. McRae (ed.) *Changing Britain: Families and Households in the 1990s* (Oxford: Oxford University Press).

Hall, C.M. and Williams, M. (2002), *Tourism and Migration: New Relationships between Production and Consumption* (Dordrecht: Kluwer).

Hampton, K. and Wellman, B. (2001), 'Long Distance Community in the Network Society', *American Behavioral Scientist* 45:3, 476–95.

Hannam, K., Sheller, M. and Urry, J. (2006), 'Editorial: Mobilities, Immobilities and Moorings', *Mobilities* 1:1, 1–22.

Harvey, D. (1989), *The Condition of Postmodernity* (Oxford: Blackwell).

Hiller, H. and Franz, T. (2004), 'New Ties, Old Ties and Lost Ties: The Use of the Internet in Diaspora', *New Media and Society* 6:6, 731–52.

Horst, A.H. (2006), 'The Blessings and Burden of Communication: Cell Phones in a Jamaican Transnational Social Field', *Global Networks* 6:2, 143–59.

Kellerman, A. (2006), *Personal Mobilities* (London: Routledge).

Licoppe, C. (2004), '"Connected Presence": The Emergence of a New Repertoire for Managing Social Relationships in a Changing Communication Technoscape', *Environment and Planning D* 22, 135–56.

Licoppe, C. and Smoreda, Z. (2005), 'Are Social Networks Technologically Embedded? How Networks Are Changing Today With Changes in Communication Technology', *Social Networks* 27, 317–35.

Ling, R. (2004), *The Mobile Connection: The Cell Phone's Impact on Society* (Amsterdam: Morgan Kaufman).

Lovelock, J. (2006), *The Revenge of Gaia* (London: Allen Lane).

Mason, J. (2004), 'Managing Kinship over Long Distances: The Significance of "The Visit"', *Social Policy and Society* 3:4, 421–9.

Massey, D. (1994), *Space, Place and Gender* (Cambridge: Polity).

Miller, D. and Slater, D. (2000), *The Internet* (Oxford: Berg).

Panagakos, N.A. and Horst, H. (2006), 'Return to Cyberia: Technology and The Social Worlds of Transnational Migrants', *Global Networks* 6:2, 109–24.

Parban, A.A. (2004), 'Diaspora, Community and Communication: Internet Use in Transitional Haiti', *Global Networks* 4:2, 199–217.

Parreñas, R. (2005), 'Long Distance Intimacy: Class, Gender and Intergenerational Relations between Mothers and Children in Filipino Transnational Families', *Global Networks* 5:4, 317–36.

Putnam, D.R. (2000), *Bowling Alone: The Collapse and Revival of American Community* (New York: Simon & Schuster).

Salaff, J., Fong, E. and Siu-Ling, W. (1999), 'Using Social Networks to Exit Hong Kong', in Wellman (ed.).

Sutton, R.C. (2004), 'Celebrating Ourselves: the Family Reunion Rituals of African Caribbean Transnational Families', *Global Networks* 4:3, 243–58.

Tanabe, M., van den Besselaar, P. and Ishida, T. (eds) (2002), *Digital Cities II: Computational and Sociological Approaches* (Berlin: Springer).

UNDP (2004), *Human Development Report* (New York: UN).

Urry, J. (2002), 'Mobility and Proximity', *Sociology* 36:2, 255–74.

—— (2004), 'Connections', *Environment and Planning D* 22, 27–37.

—— (2007), *Mobilities* (Cambridge: Polity).

Vertovec, S. (2004), 'Cheap Calls: The Social Glue of Migrant Transnationalism', *Global Networks* 4, 219–24.

Wellman, B. (ed.) (1999), *Networks in the Global Village* (Colorado: Westview Press).

—— (2001), 'Physical Place and Cyberplace: The Rise of Personalised Networking', *International Journal of Urban and Regional Research* 25:2, 227–52.

—— (2002), 'Little Boxes, Glocalization, and Networked Individualism', in Tanabe et al. (ed.).

Wellman, B., Hogan, B., Berg, K., Boase, J., Carrasco, A.J., Côté, R., Kayahara, Kennedy, J.T.L., M, L.T. and Tran, P. (2005), 'Connected Lives: The Project', in P. Purcell (ed.) *Networked Neighourhoods* (Berlin: Springer).

Wilding, R. (2006), '"Virtual Intimacies?" Families Communicating Across Transnational Contexts', *Global Networks* 6:2, 125–42.

Williams, M.A. and Hall, M.C. (2000), 'Tourism and Migration: New Relationships between Production and Consumption', *Tourism Geographies* 2:1, 5–27.

Williams, M.A., King, R., Warnes, A. and Patterson, G. (2000), 'Tourism and International Retirement Migration: New Forms of an Old Relationship in Southern Europe', *Tourism Geographies* 2:1, 28–49.

Zook, M., Dodge, M., Aoyama, Y. and Townsend, A. (2004), 'New Digital Geographies: Information, Communication, and Place', in S. Brunn, S.L. Cutter and J.W. Harrington (eds), *Geography and Technology* (Norwell: Kluwer Academic Publishers).

Chapter 9

Young Refugees in a Network Society

Marianne Brekke

Introduction

In this chapter I shall explore the way in which young refugees[1] create and maintain transnational relations and networks, and how these processes influence their everyday lives. The group of young people I am focusing on live in Tromsø, a town in Northern Norway. They are linked to different transnational networks, which have different levels of meaning in their lives. Issues discussed include the kind of meaning this northern place has for young refugees and whether they perceive Tromsø as just a temporary place or as a place to stay in the future. Interaction between young Norwegians and young refugees is limited, and a lack of commitment or sense of belonging to the place makes it even easier to perceive ideas about mobility. Many of the young refugees are already a part of a migration process and have ideas of moving back to their country of origin in the future. Before they came to Norway, many of them had to migrate from their own country and spent time in different African countries in refugee camps.

Theoretical approaches in this chapter are based on cross-cultural perspectives on young people, place and migration, which have been discussed by, amongst others, Karen Fog Olwig (2005). I shall further focus on transnational practices and networks in a migration context, following the perspectives of Steven Vertovec (2004, 2006).

This chapter is based on my PhD research on young refugees in Tromsø, in which I used qualitative research methods such as conversations and in-depth interviews. I also visited the homes of refugee families in Tromsø. Encounters where I met young refugees took place in occasional locations in town: on the bus, in the streets, in the grocery store and in the library. These brief encounters provided valuable knowledge of the kind of locations and arenas of urban space young refugees are moving between.

---

1   Those referred to using this term have either come to Norway with the status of UN refugees or they have been reunited with one or more family members who have the status of refugees. Since all of them are either refugees or come from families with a refugee background I shall refer to them as young refugee or refugees in this chapter.

Young People and Migration – Theoretical Approaches

Research on migration and young people is turning into an extensive field. Young people have been ascribed a mediating role in the family in the process of integration. By going to school in the receiving country they are supposed to learn social norms, forms of behaviour and cultural values (Fog Olwig 2005, 10). Migration in an anthropological perspective has been engaged in issues regarding networks or social communities, often based on kinship. People do not migrate as individuals but as members of a family or kinship group. Migrants, and refugees in these relationships, have close connections to, and enter into, socio-cultural communities in their native countries. In order to grasp migrant young people's perspectives, it is important to study how they construct everyday life in an interaction with their surroundings (Fog Olwig 2005).

Fog Olwig (2005) claims that immigration does not necessarily lead to integration in the receiving countries. However, does this mean that extended transnational networks result in lower levels of integration and inclusion in the new society? Young people establish different social bonds and enter into cultural relationships of belonging in their transnational as well as their local environment. We may expect migrants to develop a kind of temporary life, consisting of different senses of belonging to various places and cultural forms, which change contextually (Fog Olwig 2005). These processes have also been studied as transnationalism (see also Chapter 10 by Skaptadóttir and Wojtynska).

Steven Vertovec (2004) claims that the concept of transnationalism has been over-used in academic work, as well as being used interchangeably with concepts such as 'international', 'multinational', 'global' and 'diasporic'. Both research and theory have problematized the difference between trans-*national*, trans-*state* and trans-*local* processes and phenomena. However, Vertovec (2004, 3) defines transnationalism as 'a set of sustained long-distance, border crossing connections'. One interesting aspect that Vertovec raises is 'trans' what, and in what spheres of life transnational connections influence migrants. In studying migration and transnational connections among migrants, it is necessary to focus on aspects such as family, religion, ethnicity and identity, and remittances (Vertovec 2004).

Refugees and Locals in the Town Landscape

Every year between approximately seventy and ninety refugees arrive in Tromsø.[2] Young people involved in this study have come from Eritrea, Rwanda and Congo, and all of them have strong Christian beliefs. They have not chosen to come to Norway or to Northern Norway, but have been settled by the Norwegian authorities, or they have been reunited with family members.

---

2   Asylum seekers and people who are reunited with their families are also included in this category (email communication with 'Flyktningtjenesten' [Refugee Services] in Tromsø municipality, 2006).

In the municipality of Tromsø there are people with a refugee background from more than 45 countries.[3] The entire municipality is populated with people from approximately 130 countries. This includes immigrants from both Western and non-Western countries. Individuals from Sweden, Russia and Finland form the largest immigrant groups in Tromsø.[4]

The young people who are the focus of this study are approximately 17–20 years old. They all participate in an induction programme, learning the Norwegian language and acquiring knowledge of Norwegian society, which is both a civil right and a duty for refugees arriving in Norway. They have to complete the courses if they want to retain their permission to stay in the country, and also if they want to apply for Norwegian citizenship.[5] Because of the obligatory courses, many choose to stay in Tromsø until these obligations are completed.

Through observations and conversations with refugees and Norwegians, including people who work with refugees, it would appear that there is a low level of interaction and few encounters between refugees and Norwegian locals in the town. It is difficult to know whether refugees and locals have strategies of avoidance concerning potential encounters. In the open public spaces, however, it seems as though Tromsø is an open and inclusive town, with a diverse mix of locals, refugees, tourists, migrant workers and other visitors who intermingle effortlessly. Most of the refugees in Tromsø come from non-Western countries and stand out in the town landscape because of their skin colour. Skin colour, and in some cases the way people dress, are the only visible marks of 'otherness'. In popular understanding, multiculturalism is often confused with skin colour.

Since the town centre is quite small geographically, there are no typical streets or public areas that are occupied by only one category of people, either refugees or Norwegians. This is also the case with shopping malls and the library, which are used by every population category in Tromsø. The library is situated in the central area of town. It is easily accessible for all Tromsø citizens and has a very special function as a meeting-place. I conducted interviews with young refugees in the library because they explained that this was a place where they felt comfortable. The library is one of the public arenas where aspects such as ethnicity, skin colour, age, sex and class seem irrelevant. Refugees occupy the same physical space as white, middle-aged professors, sitting beside one another and reading books and newspapers. This kind of encounter might stand as an example for what Goffman (1961) calls unfocused interaction, which happens in interpersonal communication when people are in each other's presence without any verbal communication. Instead, people modify their behaviour because they know they are being observed. On the other hand, visitors to the library may modify their behaviour according to the library's codes of conduct. In this case it means being silent or talking quietly.

Going to the library does not involve knowing other people or embarking on any form of interaction, and people do not stare at you if you sit alone. The library

---

3   *Bosettingsstatistikk kommunevis – nasjonsfordeling i år asylsøkere fra 1995–2005* (IMDi) Integrerings- og mangfoldssdirektoratet [Directorate of Integration and Diversity].

4   http://www.tromsointerinfo.no/?page_id=10, accessed 7 May 2007.

5   http://www.imdi.no/templates/Tema____4529.aspx, accessed 4 May 2007.

constitutes a kind of 'imagined community' where the visitors feel included, even though they do not know each other. The library is also a place where transnational networks are maintained. Refugees use the Internet, chatting with people around the world, and read international publications, as well as papers from their native countries. They meet up with other refugees from different countries and these activities bring a transnational dimension to the place.

The library does not involve the need to spend money and this may be a reason why both Norwegian locals and refugees hang out there. When it comes to other semi-public places, such as shops and cafés, segregation between Norwegian locals and refugees becomes more visible. In the most popular urban coffee bars there are few refugees, and almost no refugee women. These places are dominated by white people of different age groups.

Frønes and Brusdal (2000) claim that urbanity is particularly associated with a young lifestyle that includes a dynamic life, for example, café- and night-life. Special patterns of consumption and knowledge of cultural distinctions are also important characteristics of urban life. However, multi-ethnicity is another dimension describing cities and urbanity that is not necessarily in accordance with the urbanity focusing on trendy cafés, shopping and cultural distinction.

How do the young refugees use the urban space in Tromsø? Many do not go to cafés or 'hang out' in town, but go straight home after school. However, all of them are regular users of the library. Kenny says:

> I come into town only if I have an appointment with someone, otherwise I do not hang out in town. I do not go out so much, maybe sometimes with my mum, for doing some shopping and just to look around. I am thinking if I got (Norwegian) friends here in Norway. […] Maybe they like going in cafés. Me, in my native country, I haven't experienced such things. I am not used to going out, and I feel a little insecure, and maybe I will get problems doing that.

He adds that in his eighteen years of life he has mostly been in school or at home, and says that he has inherited this way of living from his father. Influenced by life in his native country, he has brought these ideas with him to Tromsø. In some ways, the behaviour that was regarded as proper in his native country is more complex and questioning of Western youth culture.

Place, Ambivalence and Mobility

Doreen Massey (1991) claims that people are linked by networks and relations, and each 'place' may be viewed as a particular, unique point of intersection. Place, in this understanding, is seen as a meeting-place. This is based on the idea that places are imagined as 'articulated moments in networks of social relations and understandings' (Massey 1991, 28). Massey emphasizes that places are not surrounded by boundaries but are open, and relations are stretched out in space.

Young people with a refugee background in Tromsø find that they have little sense of commitment to Tromsø. However, Tromsø may function as an intersection for their social networks and their transnational relations. They have not grown up in the town and they do not know the streets, buildings and 'symbolic' locations in the way that young Norwegians in Tromsø do. The refugee parents have even less local knowledge than their children, mostly because they do not move around in the same way as their children. The parents, as well as the young people, do not have a sense of affiliation with the place, which may mean that they do not have a strong feeling of belonging. Tromsø, for them, appears to be just another intersection in a network of relations and a 'place in transit' in an ongoing migration process. The young people's future plans are in many ways connected to places other than Tromsø. It seems as though many refugees experience Tromsø as only a temporary place 'in-between' other things. In this case, refugees who are part of a migration life cycle are less interested in investing time and engagement in places with which they have no connection.

Places are not only experienced as open and boundless when seen from the refugees' point of view. When experiences of Tromsø are discussed, the physical and material environment is initially emphasized. The Arctic north offers a very harsh and cold climate and many feel isolated by staying in their homes and not keeping in touch with their family and their social network in other places on a daily basis.

Tromsø is to a large extent associated with climate and weather. The young refugees were shown pictures of Norwegian winter landscapes while staying in refugee camps in Africa. Since she never had seen snow before, a young woman from Congo anticipated that the country was covered with 'white soil'. Images and pictures of Norway formed the reality that they experienced when they arrived in town. Kenny describes his first encounter with Tromsø: 'It was polar nights, and a lot of snow, and the first time I saw snow. It was so cold that we needed a lot of clothes.' Young refugees emphasize that the harsh climate also influences the social dynamics in the town landscape. They observe that during winter there are hardly any people moving around in town and they never see people outside their houses. Migrant workers in Iceland also regard the rough climate as a negative aspect of place (see Chapter 10 by Skaptadóttir and Wojtynska). A climate that differs considerably from what people are used to seems to play an important role in how both working migrants and refugees experience a place.

Statistics show that most of the people with a refugee background who are settled in Northern Norway move to places further south in the country after just a short stay in the north (Østby 2004). One of the explanations for why they move from Northern Norway to the south is based on their experience with the harsh climate. Being part of an extended ethnic and religious community in other places, and better job opportunities, are also general explanations. Today, the social and economic costs of moving to another place are low. By chance I met an informant, Emmanuel from Rwanda, and his family at the airport in Tromsø. Emmanuel told me that his parents were moving further south in Norway, where they had an extended social network and where the climate was milder. Emmanuel and his brothers and sisters were supposed to stay in Tromsø for at least one more year to complete Norwegian courses at school. In this instance, a meaning of place also entails certain obligations, such as completing Norwegian courses.

Tromsø – an open and multicultural town?

During the nineteenth century, Tromsø was called the 'Paris of the North' for many reasons. One was because of its 'exotic urban life with a cultural diversity and impulses from the continent' (*Kommuneplan for Tromsø* 1999, 68). Fashions from larger cities in Europe, such as Paris, were imported and used by women in the town, and there have always been a large number of cafés and a lively night-life (Knutsen 1996).

Historically, the region of which Tromsø forms part has been populated by different ethnic groups from northern areas, including the Sámi, Finns and Norwegians. In recent years, immigration from other parts of the world has given Tromsø a more complex population. In latter years, Tromsø has been declared a multicultural town and an anti-racist zone. Institutions such as the University of Tromsø highlight an extended international involvement and co-operation with countries in the southern part of the world, as well as being central to discussions and research relating to Arctic areas. Tromsø seems to want to appear to be a town with major international involvement, and this is also evident in the way that Tromsø is simultaneously regarded as a unique and a global location (see Chapter 18 by Kraft).

'Open', 'inclusive' and 'hospitable' are concepts used to describe the town in public documents (*Kommuneplan for Tromsø* 1999). Politicians and a general understanding among Norwegian locals also support this view. In addition, Tromsø has a vision of being the best and most inclusive municipality in the country for children and young people to grow up in, irrespective of their language, health and social, economic and cultural background.[6] I would argue that there are many paradoxes in the way that Tromsø attempts to appear and be understood and the way in which different groups in the town actually experience it as a place.

Young refugees do not emphasize qualities such as openness, inclusiveness and international as their experience of the town. Concepts such as 'the other' and 'stranger' are more descriptive of how many of them experience their position in relation to the majority population in many situations. The different backgrounds and sense of place experienced by Norwegian locals and refugees reflect the fact that the place has multiple identities (Massey 1991).

As I have mentioned, there is a low level of interaction between ethnic Norwegian and young refugees. Refugees state that there are many reasons for this. 'I feel it is harder to get in touch with Norwegians than with foreigners […] and they (Norwegians) are very quiet', says Julia. They also emphasize that limited language skills form another major difficulty, reducing social contact with young Norwegians. Conflicting cultural values and understandings also create ambivalence about getting involved in activities and social interaction with the young majority. Kenny explains that religion has a substantial space in his life and he realizes that the Western youth culture represents ideas and practices that are not consistent with his beliefs. Going to parties where people drink alcohol, or to youth clubs where everyone dances, places him in a moral dilemma that he finds difficult to cope with. He realizes that his strong faith excludes him from arenas where he might have had the chance to get to know young Norwegians.

---

6   http://www.tromso.kommune.no/asset/17351/1/17351_1.pdf, accessed 7 May 2007.

One common understanding among young refugees is that they find Norwegian patterns of behaviour very strange. They are especially amazed by the absence of greeting in their daily encounters with Norwegians and find them very impolite in such situations.

> (When I arrived in Tromsø) people were completely strange, because every morning I used to greet people I met outside my house, but they did not respond [...] I wondered why they did not say anything, because back in Africa you greet every person you meet [...] After that, I stopped greeting. (Emmanuel)

Despite ambivalent and complex experiences with Norwegians, some of the refugees have clear sense of why they are in Tromsø. They also express satisfaction at the idea of staying in this northern town and education potential is one reason for this. Other reasons include meeting other young people from different countries who are from a refugee background, some of whom have been reunited with family members. This means that places also establish and connect relations, rather than just stretching them out. Maria explains one of her main intentions in coming to Tromsø:

> My aim is just to study, and that's why I came to Norway. Well, I came here to live, too. But, it is better to be in your own society, and speak your own language. But, because there are better educational possibilities here, I came here to learn.

Learning the Norwegian language is an ambivalent task. On one hand, learning the language offers the opportunity to establish relations with young Norwegians and the possibility of getting a job. Some informants also emphasize the fact that it is convenient to know Norwegian in a multi-ethnic situation. On the other hand, spending time learning the language is not that motivating because it is waste of time, according to other refugees.

Local and Transnational Relations and Networks

Young refugees have strong links and relations with societies other than their temporary place of living, which for the moment is Tromsø (see Fog Olwig 2005). As Vertovec (2004) discusses, it is interesting to map the spheres of life in which refugees are involved in transnational networks. One aspect that Vertovec (2004) stresses in this definition is that transnationalism is a connection that crosses borders. Border is a vague concept, and since some connections are maintained mainly through technology, the experience of crossing national borders disappears. For instance, chatting on the Internet rarely gives a sense of crossing national borders, even if people are of different nationalities. The concept of trans-*national*, then, is in many senses inadequate to describe these connections. Even though other concepts such as trans-*local* or trans-*state* might replace it, this will not provide any more of an adequate understanding of the phenomenon (Vertovec 2004).

Crossing national borders or not, young refugees in Tromsø engage in long-distance connections that have an impact on their daily lives. They spend a lot of time chatting on the Internet with friends located in countries other than Norway.

Some have jobs and part of their wage is sent as a remittance to family members in their native countries. Some of these practices demand simultaneous interaction by the individuals involved in the different places, like chatting on the Internet and phone calls, while sending remittances does not require simultaneity.

Refugees' homes in Tromsø provide a place to maintain family life as well as both transnational relations and local relations with friends they have met in Tromsø. Visiting an Eritrean home provided an image of everyday life in a refugee family and the meaning of transnational relations, as well as local networks. From this visit I obtained an insight into how the dynamic life in the family takes place. In the living room two personal computers are located very centrally and these are in constant use. Two televisions are on very loud, one in the kitchen area and one in the living room, showing American soaps subtitled in Norwegian. They are cooking Eritrean food and the apartment is filled with cooking smells that are unknown in Norwegian homes. Friends of a refugee background from Afghanistan and Rwanda are also present. Different languages are being spoken in this multi-ethnic setting, including local Eritrean languages, English and Norwegian. The Eritrean language is spoken only between family members and they switch easily between the different languages, depending on whom they are talking to. Everyone is required to master Norwegian to a certain level, since it is the only common language that all the young refugees know in this setting. They need to know the language of the majority population when they are communicating with friends of different nationalities, even in a setting where there are no Norwegians.

Friendships develop across ethnic and religious boundaries and the young people stress that these aspects do not play any role in making friends. When Christian Eritreans intermingle with Muslim Afghanis, religion and ethnicity is never an issue. Inter-ethnic friendships form an important aspect of refugees' everyday life in Tromsø. However, there are very few ethnic Norwegian young people involved in these mixed friendships. The open-minded attitude to ethnicity in friendship constructions among young people from a refugee background has also been demonstrated in a study conducted in both Oslo and Tromsø (Engebrigtsen and Fuglerud 2007). The presence of refugees from many countries in Tromsø results in communities made up of people of different ethnicities and nationalities. I would argue that this diversity has had many positive effects. In multi-ethnic communities, negotiating cultural and social practices is necessary if social interaction is to happen. Viggo Vestel (2003) demonstrates how people in multicultural contexts emphasize aspects that are common to groups, instead of emphasizing differences. However, in order to make interaction possible it is crucial not to suppress important differences between groups by only celebrating homogeneity and sameness, nor to place an extreme focus on diversity to the extent that interaction is not possible. Vestel introduces the condition in-between these two approaches as cultural elasticity. Interaction and encounters between young refugees are based on what they have in common, their experiences as young people of a refugee background and their ambivalent relationship to the majority society. This common background creates a social community that is more important than emphasizing different ethnicities and religions. Language adaptation in social situations is also an important aspect and is one example of cultural elasticity and flexibility.

Online – transnational relations

Chatting on the Internet is the most important way of keeping up with 'online friends' all over the world. Many do not maintain connections with their homeland via the Internet because the Internet connection is not very well developed in the rural areas where some of the refugees come from. However, many of the young people have met new people from other countries on the Internet, and some keep in touch with other refugees from their homeland who live in other parts of Norway. Maria from Eritrea has met a friend in a chat room on the Internet. She tells me that their common interest in religious issues has brought them together. At certain times during the day they arrange a meeting in a private chat room. Maria says:

> I started with chatting because I had no friends here…and I wanted to spend my time doing something. It is a boy that I chatted with from Kenya. He started to talk about God. I want to hear about God from other people…So we chatted about God, about jokes. We exchange songs. I give him songs from my native country, and he gives me English Christian music. We ring each other sometimes.

Next time I meet her she tells me that they consider their relation to be a love relationship. They want to visit each other, but she admits that this might be complicated. She tells me that members of her family, both in Tromsø and in Eritrea, have become involved and know about her online relationship. Since it is a very important tradition in her native country that the family accept the partner, her grandfather has met her new boyfriend in Kenya. He wanted to ensure that this young man was suitable as a potential future member of the family. This example shows how young people meet via technological means of communication, but it also illustrates that being accepted in a family still requires traditional customs, like meeting people in face-to-face situations. It demonstrates co-operation between family members, intersecting cross-generational and traditional aspects, as well as the important meaning of transnational relations.

Maria explains that she started chatting because she did not have any friends. Zygmunt Bauman (2001) claims that having access to cyberspace has resulted in less autonomy in local communities, and transmission of information and relations on the Internet has contributed to a devaluation of place (Bauman 2001). For many young refugees, the Internet provides a way out of loneliness and marginalization when they arrive in Tromsø. By using the Internet they are not so dependent upon finding friends and developing social networks in their geographical proximity, and regard online friends as being just as adequate as the people they meet face-to-face. However, places are also meeting-places where social encounters occur between people. Tromsø as a place for refugees seems, then, to have multiple meanings, both as a place where transnational connections are maintained but also where social encounters are established.

Commitment and interdependency in transnational families

Some of the young refugees live with only one parent, while the other parent lives in their homeland. Though the family is spread geographically, they feel interdependent. The impact of migrant transnationalism and practices has an emotional cost (Vertovec 2004). Young people tell me that they miss their family members very much in daily life and it is very hard to accept the uncertainty of when they will meet them again. In my study I met a group of three brothers and sisters who live with their mother in Tromsø. However, they also have strong connections to their father in their native country and call him regularly. The telephone is the most important means of communication with regard to maintaining relations with family and friends in the native country. Dramatic reductions in phone costs have resulted in the ability to phone family members on a regular basis. Phone calls create personal, real-time contact that has transformed everyday life for migrants all over the world (Vertovec 2004). Extended connections between migrants and their families via phone calls, and the way in which the use of text messaging has become part of long-distance parenthood is discussed by Larsen and Urry in Chapter 8.

Other forms of connection, such as sending remittances, are more institutionalized. Some of the young refugees have jobs, as well as going to school. These are jobs with cleaning firms and global fast food chains located in Tromsø. In the case of the family mentioned above, the migrant young people send remittances to their father. By sending remittances, an interdependent relation has been established between the young people and their parents. The young people fulfil their parents' expectations and family members in their homeland are able to improve their standard of living.

As I have discussed, the refugees demonstrate great competence in using different means of communications, such as phone calls, the Internet and sending remittances in order to maintain relations with friends and families who do not live in their physical proximity. They display a knowledge of how to maintain social relations at a distance. These mobility processes could also be understood as network capital (see Chapter 8).

Limited Circular Migration?

For many migrants today, migration is part of their life cycle. Circular migration describes a process whereby migrants, through their movements, extend and establish networks connecting places of origin and places abroad (Vertovec 2006).

Many immigrants who came to Norway during the 1960s and 1970s as legal working immigrants have the option of travelling back and forth between their native country and Norway whenever they want to, as is also the case for immigrants of Pakistani background and their descendants. Refugees do not have the same option of going back to their native countries because of ongoing wars and conflicts in the regions from which they fled. They do not have the same rights and options as people who have acquired Norwegian citizenship. This means that they do not know when they can ever go back. However, all of those involved in my study have the ambition of one day returning to their place of origin. Young refugees are already in an ongoing, circuit-based migration process, both when it comes to geographical movement and with regard to educational mobility. As I have already mentioned, they stayed in refugee camps in African countries other than their native country before they arrived in Tromsø. The majority of migrants of refugee background will stay in Northern Norway for only a temporary period before they continue their migration process to places further south. Going to the United States is a dream for quite a few, because many have heard that they will achieve the best education and work potential there.

> I don't know anything about the universities here, but I think they could be more developed in the USA [...] Everybody there comes from all over the world. [...] I think that America has developed more, a higher income, from what I learned in my country. (Maria)

Maria also stresses the point that the USA has a mixed population, with its roots from all over the world. I interpret this as her understanding that she will feel more included in such a society than in Norway, which is considered to be culturally very homogeneous. Although the status and rights of a refugee provide limited mobility, this does not prevent young people from having ambitions and dreaming of migration.

Young Refugees: In-between Places and Networks

In this chapter I have discussed a number of aspects of young refugees' everyday lives, as well as transnational practices and their impact. Tromsø is perceived as a place relating to ambivalent experiences and it appears to be more or less a 'place in transit', seen from the refugees' perspective. The paradox is that the town considers itself, and wishes to appear, multicultural and open, a place where people from all kinds of backgrounds can feel included. Many refugees perceive themselves as being in an ongoing migration process and do not identify with the place, which quite a few consider a temporary place in which to live. However, Tromsø matters to them, both as a place where new relations are established and as a point of intersection for transnational connections. Young refugees have very few connections with the majority population. The general experience is that they find it difficult to get to know young Norwegians. Participation, in Norwegian youth contexts, is restricted by conflicting ideas and dilemmas relating to religious, cultural and social practices. Refugees' attitudes and practices might also be analysed as a kind of escapism from a new society, as they search for communities and people who have more or less the same background as themselves. However, the library in town is a place visited by both refugees and Norwegians. Face-to-face interaction is not necessary, but it is a place where transnational networks are maintained, mainly via online relations.

The Internet, phone calls and remittances are ways of being connected to families, relatives and friends who belong to different communities and places outside Tromsø. Networks, connections and commitments elsewhere make ideas of mobility more feasible and realizable in these young peoples' lives.

References

Bauman, Z. (2001), *The Individualized Society* (Oxford: Polity).

Engebrigtsen, A. and Fuglerud, Ø. (2007), *Ungdom i flyktningfamilier. Familie og vennskap – trygghet og frihet?* NOVA Rapport 3/07.

Fog Olwig, K. et al. (2005), 'Små og store fortællinger: Børn og unge i migrationsforskning', in L. Gilliam et al. (eds).

Frønes, I. and Brusdal, R. (2001), *På sporet av den nye tid – Kulturelle varsler for en nær framtid* (Bergen: Fagbokforlaget).

Gilliam, L. et al. (eds) (2005), *Lokale liv – fjerne forbindelser. Studier af børn, unge og migration* (København: Hans Reitzels Forlag).

Goffman, E. (1961), *Encounters – Two Studies in the Sociology of Interaction* (Indianapolis, New York: The Bobbs Merrill Company, Inc.).

Hodne, B. et al. (eds) (2003), *Kulturforskning* (Oslo: Universitetsforlaget).

IMDi [Integrerings- og mangfoldsdirektoratet, Directorate of Integration and Diversity] (2006), *Bosettingsstatistikk kommunevis – nasjonsfordeling i år fra 1995–2005.*

IMDi [Directorate of Integration and Diversity], '*Norsk og samfunnskunnskap*', http://www.imdi.no/templates/Tema____4529.aspx, accessed 4 May 2007.

InterInfo, *Tromsø Kommune, 'Befolkningen i Tromsø kommune etter statsborgerskap 1. januar 2005 og 1. januar 2006'.* Statistics page, http://www.tromsointerinfo. no/? page_id=10, accessed 7 May 2007.

Kommuneplan for Tromsø kommune (1999), [Municipality Plan, Tromsø municipality].

Knutsen, N.M. (1996), *Nordens Paris. Vandringer i Tromsøs muntre liv og historie* (Tromsø: Angelica).

Massey, D. (1991), 'A Global Sense of Place', *Marxism Today*, 24 May–9 June.

Prieur, A. (2004), *Balansekunstnere. Betydningen av innvandrerbakgrunn i Norge* (Oslo: Pax Forlag A/S).

Program for inkludering og godt oppvekstmiljø for minoritetsspråklige barn og ungdom i Tromsø kommune (2007), http://www.tromso.kommune.no/ asset/17351/1/1735 1_1.pdf, accessed 7 May 2007.

Simonsen, K. (2005), *Byens mange ansigter – konstruktion av byen i prasis og fortælling* (Frederiksberg: Roskilde Universitetsforlag).

Vertovec, S. (2004), *Trends and Impacts of Migrant Transnationalism*, Working Paper No. 3, Centre on Migration, Policy and Society (COMPAS).

—— (2006), 'Is Circular Migration the Way Forward in Global Policy?' in *Focus on Skilled Migration, Around the Globe* (Monash: Institute for the Study of Global Movements) 3:2, 40–4.

Vestel, V. (2003), 'Et fellesskap i gråsonen' in Bjarne Hodne et al. (eds).

Østby, L. (ed.) (2004), *Innvandrere i Norge – Hvem er de, og hvordan går det med dem?* Del 1, Demografi, Notat, Statistics Norway 2004/65.

Chapter 10

Labour Migrants Negotiating Places and Engagements

Unnur Dís Skaptadóttir and Anna Wojtynska

Introduction

Global and transnational perspectives are increasingly challenging earlier conceptions of places as stable culture units inhabited by people who share a common identity and are bound by solidarity. In times of *liquid modernity* it is more difficult than before to sustain such a stance (Bauman 2000). There is a growing literature that deals with the complex relationship between culture, place and belonging (Appadurai 1996; Bauman 2000; Gupta and Ferguson 1997a; Gupta and Ferguson 1997b; Hannerz 1997; Kearny 1995; Savage et al. 2005; Urry 2001). Discussing globalization, Inda and Rosaldo, for example, argue that in the current world in motion 'people readily cut across national boundaries, turning countless territories into spaces where various cultures converge, clash and struggle with each other' (Inda and Rosaldo 2002, 3). At the same time, places continue to be significant for people's sense of identity and narratives of belonging if, as Escobar explains, 'we understand by place the experience of a particular location with some measure of groundedness (however, unstable), sense of boundaries (however, permeable), and connection to everyday life, even if its identity is constructed, traversed by power, and never fixed' (Escobar 2001, 140). In this chapter we examine how people who have come from other countries to work in fishing villages in a remote area in the north-west part of Iceland experience the places where they are living. The examples drawn upon in our discussion are from our studies during the last decade of the twentieth century and the beginning of the twenty-first century.[1] We have used qualitative research methods that include interviews, informal discussions and observations in an attempt to analyse and understand people's perceptions and experiences. We begin with a presentation of the theoretical perspectives underpinning our discussion. We then examine the development of labour migration to these fishing villages, followed by an exploration of the immigrants' views of their new locales. Since Poles are by far the largest group of foreign-born residents in the area, our examples primarily focus on their views and experiences. As we show, places continue to be vital, in terms of both their former and current homes, although the meanings of these places are continuously contextualized and negotiated through the transnational migrant experience.

---

1   Unnur Dís Skaptadóttir has conducted ongoing research in the area, with five different research trips since the early 1990s, and continues to follow developments in the region. Anna Wojtynska conducted field research in the area for her doctoral research in 2004.

Theoretical Perspectives

Growing attention to global processes since the 1990s has to a large extent shifted the focus of immigrant studies from concerns with integration and ethnicity towards considerations of mobility, transnationalism, translocality and transmigration. These concepts are used to draw attention to mobility in the analysis of the multiple attachments migrants may have with more than one nation or locality (Basch et al. 1994; Castells 1996; Vertovec 1999; Vertovec 2004). Researchers have shown that many people migrate not only in order to begin new lives for themselves, but just as much to improve the lives of their families back home (Sørensen and Olwig 2002). The economy of many households relies to a large extent on the mobility of its members. Sørensen and Olwig call this type of migration mobile livelihoods and they 'advocate shifting the analytical focus from place to mobility, and from 'place of origin' and 'place of destination' to the movements involved in sustaining a livelihood' (Sørensen and Olwig 2002, 2).

The transnational lives of contemporary migrants challenge, as Levitt and Glick-Schiller argue, 'our understanding of basic social institutions such as the family, citizenship, and the nation state' (2004, 3). Consequently, transnational research implies a re-conceptualization of the notion of society. Our analytical lens must broaden and deepen because individuals are embedded in multi-layered, multi-sited transnational social fields, encompassing those who move and those who stay behind (Levitt and Glick-Shiller 2004, 3). Diverse aspects of transnational experiences have been studied, such as continued contact with the country of origin, economic remittances and political or cultural involvement in the home-town (Basch et al. 1994; Guarnizo 2003; Parreñas 2001). Based on growing research on transnational ties, Wimmer and Glick-Schiller (2003) criticize what they call 'methodological nationalism', which takes the nation-state as a given research unit, and point out that although nation-states remain important, social life is not confined to the boundaries of the state.

Thus, many studies currently focus on mobility and cross-border activities, and how these activities formulate places and societies. Amongst other things, researchers are concerned with how participation in a life lived in two or more places affects the identity and sense of belonging. Vertovec, for example, in discussing the lives of migrants, points out that 'many migrants today intensively conduct activities and maintain substantial commitments that link them with significant others (such as kin, co-villagers, political comrades, fellow members of religious groups) who dwell in nation-states other than those in which the migrants themselves reside' (Vertovec 2004, 970–71). Amongst individual migrants living transnational lives this entails the emergence of a dual orientation or bifocality in everyday life. Vertovec points out that 'the transformation of everyday orientations concurrently toward both here and there is a mode of change that accompanies the transnationalization of distinct social practices and institutions among migrants' (Vertovec 2004, 976). Bifocality, he says, is something that cannot easily be measured 'but its workings are clearly discernable in social practices and conveyed in individual narratives' (Vertovec 2004, 977); these practices have considerable impact on 'individual and family life course and strategies, individuals' sense of self and collective belonging, the ordering of personal and group memories, patterns of consumption, collective socio-cultural practices, and approaches to child-rearing and other modes of cultural reproduction' (Vertovec 2004, 977). Migrants thus participate in a new society while simultaneously maintaining ties with their places of origin. Now, as never before, migrants can sustain and act upon a strong sense of connection with people, places and a sense of belonging connected with their place of origin (Vertovec 2004). Sørensen and Olwig (2002) similarly point out that the expansion of the space for personal or familial livelihood practices to two or more localities involves, in many cases, not displacement but rather multiplacement.

According to Olwig (2003), it is in many cases more useful to see movements as occurring within family networks, rather than population movements between nation-states. She argues that '[...] migrants do not only construct or perceive places of origin in terms of their transnational character, but just as much in terms of particular cultural values and social ties that the migrants, and their families, practice in relation to these places' (2003, 788). This is important in relation to our discussion of the transnational sense of belonging of migrants in Icelandic villages, where family ties often continue to play an important role in migrant experiences.

The villages where the research discussed in this chapter was conducted have never been stable places, and the mobility of people and things has always been important to their existence.[2] Seasonal workers and fishermen who leave to fish elsewhere have always been important in these villages. However, the current changes are of a different nature to earlier ones, as these villages have clearly become multicultural, especially since many of the labour migrants from abroad have settled there. These new inhabitants are negotiating their place in these localities. Today, with new technology and easier travelling, the inhabitants' activities increasingly transcend place. A transnational perspective that focuses on mobility is not only essential in order to understand the transformation of the fisheries in Iceland – the latter are very much part of the global economy – but is also important if we want to understand the daily lives of people living there.

Labour Migration and Fish Processing

Fish processing along the North Atlantic rim is increasingly a global business, not only because of international global capital and fish catches but because of a mobile workforce as well. Remote Icelandic fishing villages and towns are staging significant transformations in the forms of social and cultural encounters performed through this workforce mobility (Skaptadóttir 2004). This is certainly the case in the West Fjords, an area with one larger town of 2,734 inhabitants and smaller villages of 200–900 people. This region of Iceland had the highest proportion of foreign-born population in the country until the year 2005 (Statistics Iceland 2007).

International migration is a fairly new phenomenon in Iceland, but it is growing rapidly. In 1981, foreign citizens comprised only 1.4 per cent of the total population, whilst in 2006 this proportion had reached 6 per cent. Coastal villages, like those of

---

2   As shown for a Northern Norwegian fishing village in Chapter 7 by Gerrard.

the West Fjords, have played an important role in this development. Already by the 1970s and 1980s, temporary workers were arriving in the West Fjords from countries as diverse as Australia, New Zealand, South Africa, Sweden and Denmark. They were few in number and they were predominantly young, single individuals who stayed only for a season. At the same time, the out-migration of native-born Icelanders from the area of the West Fjords had already been an ongoing process and was the highest in the country between 1997 and 2002 (Statistics Iceland 2004). Thus despite fewer jobs, due to restructuring of the fishing industry, the fish processing firms began to suffer from a considerable lack of workers by the early 1990s. Managers actively searched for foreign labourers, resulting in new channels of population flow. In 1996, 70 per cent of all new work permits granted in Iceland were issued to firms hiring workers in fish-processing jobs (Directorate of Labour 1999).

Significantly, even though the largest numbers of immigrants reside in the capital area (and recently in the south-east corner of the country, see Chapter 17 by Benediktsson), the percentage of foreign-born residents among the population has been much higher in the coastal villages, such as those we researched in the West Fjords. In 1999, for example, foreign citizens represented 7 per cent of the population in the area, whereas in other regions this ratio ranged from 2–3 per cent. The concentration of migrants within the region varies, however, and is much higher in villages than in farming areas. There are some places where more than half the population are foreign-born or of foreign ancestry. However, the current ratio of foreigners in the population of the West Fjords remains fairly constant at around 7 per cent (Statistics Iceland 2007). Compared with other parts of Iceland that have experienced rapid increases in foreign labour, the West Fjords are becoming more stable in terms of population flow, since many of the immigrants have settled in the area.

The development of migration streams to the West Fjords, and to Iceland in general, has to a great extent been the result of personal networks. Initially, the managers of fish-processing plants used personal contacts when searching for labour, for example those which were forged when Icelandic vessels were built or renovated in Poland. These links with Poland proved to be particularly fruitful, as Poles soon outnumbered other nationalities in the West Fjords and in the Icelandic labour market in general. Presently, 62 per cent of all foreign citizens in the region are from Poland, followed by Thailand (6 per cent) and the Philippines (5 per cent) (Statistics Iceland 2007). There are also people from other countries, such as Slovakia, Russia and Portugal, and smaller numbers from South American countries.

The majority of the immigrants currently living in the West Fjords initially came for only a limited period of time. Poles in the early 1990s, for instance, commonly arrived in groups of between eight and ten individuals for half-year or one-year contracts, staying in housing provided by the firms. Since work in fish-processing was defined as a female job, there were originally more women than men among the migrants in the region. In 1995 there were twice as many women as men. Today, however, although women are still more numerous than men, the numbers are increasingly becoming more equal in gender terms. The immigrants are a diverse group, consisting of people of different educational backgrounds, ages and marital status. While many have returned to their country of origin, some of them have settled for longer periods of time. A number of people prolong their stay from year to year, and in certain cases have stayed up to ten years or more, but some still talk of returning in the undefined future. Individuals arriving on their own or with a group of workers are increasingly a thing of the past in this area, and most of the immigrants living there now reside in family units. A number of women who came to work in the fish factories were divorced or single mothers when they arrived. Some of them have married Icelandic men and subsequently brought other family members to stay with them, most commonly their children or siblings. Marriages between men who come to work as unskilled labourers and Icelandic women are rare. Because it has become more difficult to find new jobs in the region, immigration to the West Fjords is less dynamic, as we have already mentioned, and people who are currently arriving are usually related to someone already living and working in the area. Access to jobs has become competitive, which has caused some tension. For instance, we were told that those who arrived as the spouse of an Icelandic citizen were able to get work more easily, compared with those without such a connection. However, there is still a measure of fluidity to be found among the local migration patterns, due to the ongoing out-migration to the capital area on the part of both Icelandic and foreign-born residents.

With a growing number of immigrants in the area, nationality has become a new marker of difference within the villages and an important issue for the local authorities providing services to inhabitants. Questions are beginning to be raised as to whether these localities, which formerly had a strong sense of community, have to a certain extent become spaces divided into subspaces? How do people cope with those changes? How do immigrants carve out their place in the locality? Do they develop feelings of attachment to these places?

Transnational Social Space

The villages included in our research may be claimed to be multicultural, where people of different origins share everyday space at work and in the neighbourhoods. They have literally become meeting-places or a *global region*, to quote Saskia Sassen (1991). This area could be described as 'postnational zones' (Appadurai 1996; Inda 2000), spaces where people of different origins meet, compete and negotiate their place. It is no coincidence that the *Celebration of Nations* was first held in this area in 1996. This was organized by grass roots organizations and the local authorities. There is a multicultural centre providing services for immigrants locally as well as all over the country, located in Isafjörður, the largest town in the region. The welcome note in the local kindergartens is commonly in four or five languages. A Polish Catholic priest also serves the local congregation. Although the largest numbers of migrants in the villages have come to work in the fish-processing plants and at other unskilled manual jobs, some have come to work in skilled jobs such as, for example, music teaching. Most of these teachers have come from Central and Eastern Europe and play an important role not only in the music life of the region but also in the local theatre.

The West Fjords may be said to belong to a transnational social space (Faist 2000), spanning distant places through various transnational practices forged by the people living there, not only temporary migrant workers but also those who have become more settled and invested in a house or established a family in Iceland. Transnational ties are produced and reinforced by, amongst other things, travelling to visit relatives or going on holiday to their country of origin. However, although travelling is easier and cheaper than before, there is an added expenditure when living in remote places like the West Fjords. Many people whom we interviewed during the course of our research maintained that they did not travel to their countries of origin as often as they would have liked. Still, modern technology including videos, less expensive telephone calls and Internet connection allows them a simultaneous participation in two places: that of their origin and their current residence. Virtually all the people who participated in our studies maintained regular contact with their community of origin, using the telephone or the Internet, and the majority of them sent remittances to their relatives, or other forms of support. Transnational migrants' activities have had a significant impact on all the localities involved, and they affect not only mobile individuals but also other residents. A flow of commodities and ideas follows a movement of people. The perpetuation of the flow of migration is also an outcome of these links. Others back home related to these migrants or returnees perceive migration to Iceland as a possible alternative in their future. Some take advantage of having a relative or a friend in Iceland and come for a short period of time to earn additional, 'quick' money.

People are connected to a locality in their country of origin through their obligations, mainly towards family members, but also occasionally towards other members of the community. As already mentioned, some migrants work in Iceland primarily in order to provide for their families back home, but even those who have settled usually have someone they say they ought to help. This feeling of obligation also stimulates the perpetuation of streams of migration. Those working in Iceland arrange work for people to help them with their difficult situation at home, or just because they believe they would have a better life in Iceland.

The intensity and modes of transnational relations differ between individuals, however, depending upon their situation and motivation. On the one hand, those who have established families in Iceland, or perceive their residence in Iceland as more permanent, have fewer contacts with their country of origin. On the other hand, strong links with the home country are also maintained by transnational families in the case where one member stays in Iceland to provide for the family. Not many are prepared to put up with the extended periods of separation from a spouse or children, which could have a damaging effect on the marriage or family as a unit (Skaptadóttir and Wojtynska 2006). As a result, these migrants feel pressured to choose either to bring the family to Iceland or to return home. In the West Fjords we interviewed a few women who had arrived as single mothers and left their children behind with their relatives.

Dichotomized Lives

As already mentioned, work is the main rationale for the arrival of the vast majority of foreigners in the West Fjords. Work thus plays a central role which absorbs most of the migrants' daytime activities and leaves little time for anything else. Those without families in Iceland, especially, accept extra hours eagerly and commonly work more than ten hours a day. After such a long working day there is not much time to socialize or explore the area. Thus, many of those we interviewed have not developed a strong attachment to the locality and do not feel emotionally connected with the place. They have no knowledge of its history, unlike the Icelanders living there. Many of them perceive the locality merely as a physical context of work. The landscape in particular, which is usually depicted as exceptionally beautiful by Icelanders, and a source of strong feelings, is talked about as more of an obstacle by the migrants. Whereas the Icelanders describe the harsh nature as playing a key role in the formation of their strong national character, the immigrants commonly complain about the constant danger of avalanches, long and dark winters and strong winds. In fact, for some it was a reason for leaving before the completion of their contracts. Similarly, the fish-processing plant has a different meaning for foreigners than for Icelanders in the villages. Whilst it is just a working place for the migrants, who feel primarily responsible for their duties at work, most of the Icelanders see the plant as a common good that is important for the welfare of the community. This is because in the development of the villages, the fish plants and fishing vessels were potent symbols for the prosperity and viability of the villages.

It seems that for some people, life in the West Fjords is defined by work and detached from leisure. Leisure time is associated with the family and community back home. In their country of origin they spend holidays and free time, meet friends and participate in celebrations of different sorts. The localities they come from are the arenas that matter most in building social status and their perception of themselves. For many of the migrants, it is of greater concern how they are perceived by others 'back home' than in Iceland. One woman explained in an interview that in Iceland she cuts her hair herself and is not worried about the result, but when she goes home she gets her hair cut and has her highlights done. Being in Iceland is only temporary and therefore all kinds of inconveniences are bearable. For the same reason, it is not important to invest in living conditions in the current locality. As mentioned above, many Polish workers lived together in housing provided by their employer when they first arrived in the area. Despite extending their stay, many of them continue to live without incurring unnecessary expenses, such as renovating their homes or purchasing new furniture, as they are saving money for when they return. Many have already invested in an apartment or house back home and some of the flats were stamped by their temporality. A couple who had stayed for several years in the West Fjords without making the decision to settle had this to say:

Husband: This is what our house looks like – everything is just improvised. Really: I made this furniture myself, and this was restored. [...] We could be living somewhere else by now. We could afford to buy a house in Ísafjörður. We've been here for nine years, so we could have saved up the money. So we could be living at a different level, for sure. And our car would be different, and everything would be different.

Wife: The flat is furnished so that we can live nicely, but it's all just temporary.

Husband: My wife did the paintings, the shelves were made by me, and so was the furniture: that's why it's got arms and legs.

Wife: We moved into this apartment recently, but until then we'd lived in a factory-owned apartment.

Some of the participants in our studies were quite isolated from Icelandic society at the time of our research; others stated that they had been isolated when they first arrived. They remarked that they had a very unclear idea of life outside the factory or the village and their contacts had, for the most part, been with other foreigners and people from back home. They did not participate in local events and rarely travelled outside the villages. Since it was most important to travel to their country of origin during the holidays, they may not have had the time or money to travel to other parts of Iceland. One man who had been living in the West Fjords for around five years, whose wife and two children live in Poland said: 'I have heard that in Akureyri it is beautiful. I have not been there yet. I do not go there because now I am concentrating on work and sending money home.'

Similarly, their contacts with Icelanders are rather limited, mostly because of a lack of knowledge of Icelandic, and often of other languages. Since most of the people had no particular interest in the country, and knew little of Iceland before arriving, they had little interest in learning the language initially and many said that they thought it would be more useful for their future to learn English. Many of those we interviewed perceive themselves to be more like guests in Iceland and feel obliged to adhere to certain forms of behaviour. As one man said, 'I am not a quarrelsome person and I am not short-tempered. I try rather to give way. In that sense I am not at home. And, on the other hand, it is better to interact than to argue.'

Physically present and often involved in personal relationships with people living in the area (Icelanders, fellow expatriates or other immigrants), many still belong mentally elsewhere. Several Poles watch Polish television (at a certain point it became quite popular to bring satellite television from Poland), not only because of the language but also because, as one woman who had lived for several years in Iceland with her family said, 'I am more interested in what is happening in my country.' Even though the village in the West Fjords has become the place of their daily life, Iceland is not 'their country'. Although migration is extending into an undefined future, they dream of returning back: at least when they retire.

At the same time, as people slowly begin to feel relatively settled, the place holds less of a feeling of novelty than before for these migrants and stops being surprising or provoking insecurity. Gradually, people obtain the practical knowledge necessary to create their everyday lives. Again, such instrumental participation does not necessarily make them feel as though they belong. A woman who had lived in Iceland for about 15 years said: 'I know for example how to buy stocks. I know all about this. I know how to vote. But I will never feel Icelandic inside. Everything in Iceland is secondary in comparison with Poland, it is less authentic.' When asked about difficulties and what she disliked, she said: 'Everything is second. Everything is second.'

Bifocality

There is less of a circulation of migrants in the region than before, and fewer individuals are there for just a short period of time. The migrant population is more stable and consists of several families. Therefore the strong division between home in their country of origin and work in Iceland is no longer demarcated so clearly. Some people claim to have two homes, both 'here' and 'there'. Their attachment grows in relation to the number of years spent in Iceland, sometimes leading to a sort of local patriotism with regard to the village or town, but not Iceland in general. What the people we talked with called their homeland was always their country of origin. Some, however, had begun to view the village as their new home.

A feeling of belonging often relies on one's ability to build attachments. Sometimes making oneself feel comfortable needs little effort. In order to make a space more familiar some people bring different things from 'back home', such as plants and food. One Polish woman, for example, had created a garden by her house and grew vegetables from Poland. Another woman from Thailand had decorated her house with ornaments and pictures from home. In this way, these migrants maintain a bifocal view and a form of transnational habitation as their place of origin continues to be important in various ways for everyday life.

Migration and life in a new locality are constantly re-evaluated and acknowledged by reference to and comparison with the place of origin. Many of those who were interviewed are satisfied with their lives in Iceland, compared with the lives they had before migration. They do feel a sense of belonging to the village in terms of their work, and as local residents. They claim that they are engaged in similar work to Icelanders living in the villages. When a Polish woman was asked if her work had become worse or better than her job in Poland she replied: 'Well, here everyone works in fish. Very few people work in the bank – four or five people in the post office – three or four. All the rest are working in fish. They [Icelanders] work in fish.'

Moreover, the Poles are aware of the fact that they are perceived as hard-working and the preferred labourers in Iceland. In addition, and contrary to the experiences of other Poles they know who have travelled to work in other European countries, they have not had to struggle with negative stereotypes. In the small localities of the West Fjords, where everyone knows everyone else, Poles seem to suffer less prejudice than some other migrant labourers in Iceland.

The following story of a Polish man in his early thirties, who had worked in a fish plant for four years, illustrates life stretched between two places. He said that he did not have many opportunities to meet Icelanders in the area, as it was inhabited mostly by other foreigners. He worked primarily with other Poles and his working day was long. He said that in the first few years he had no interest in knowing the country he was living in and found it more useful to learn English than Icelandic. He spent his leisure time reading magazines and books from Poland and watching Polish television. Now that he has planned to stay for a longer period of time in Iceland than he originally intended, he has begun to take an interest in learning the language as well as the possibility of getting a job relating to his education. However, he has already invested in land in Poland where he intends to build a house. So, along with greater participation in the West Fjords, he is simultaneously making attempts to nurture his ties with Poland.

Discussion and Conclusion

In this chapter we have discussed the meaning of place in a transnational context and examined the interconnectedness of mobility and place as it is constituted in small villages in the north-west part of Iceland. We have shown that mobility is an important aspect of immigrant perceptions of living in the West Fjords, and of Iceland. This comprises not only mobility with regard to place but also the mobility of things and ideas. We have depicted the way in which transnational/translocal practices and networking tie migrants to two places: that of their country of origin and that of their destination. The people involved in our study, based on their evaluation of possibilities, have decided to pursue work in a new place and even settle for a period of time. They have a mobile livelihood, working in Iceland in order to create a better life back home. Since employment is the primary reason for moving to Iceland, many people view the new village more or less as a place of work. Many of them rely on the migrant community and the community back home to maintain their identity. The transnational practices of regularly travelling back home, sending remittances and keeping in touch with the home country have shaped the meaning of both places. They simultaneously participate in Icelandic society, in different forms and to different degrees, and are active in the labour market upon arrival in the country. A bifocal view, such as that suggested by Vertovec, is a common experience, in which the new place is continually evaluated and compared to the place of origin. We agree with Vertovec that bifocality is not easily measured but may be observed in the context of social practice and individual narrative. Moreover, these practices have an important effect on people's strategies, consumption patterns and diverse sense of belonging. They live in a society or a social field and have family ties crossing geographical borders. At the same time, localities continue to be important, and even gain a symbolic meaning when they exist for a large part of the year in their thoughts and discussions, experienced through modern media and communication technology. Inhabitants who come from diverse countries are connected by obligations towards family and friends left behind. This is not a simple process and their experiences are diverse. People live with the memories of the place they left, but they may not feel that they belong there anymore. In spite of the fact that people maintain day-to-day relations, as we have discussed, the lack of face-to-face contact and being distanced from local events loosens the sense of belonging. Some people express their position as betwixt and between, others as belonging in two places in different ways. The notions of 'migrant labour' and 'immigrant integration' do not adequately capture the very complex ways in which the people in our studies are situated.

The new place is given meaning, but meanings that may differ from those held by the people born in the region. These meanings are based on daily experiences but they are complex; they may be informed to a lesser extent by the surrounding areas or Iceland in general and instead informed by the knowledge and views that these migrants bring with them, which are incorporated into the new setting. Thus, 'place' is continually reconstituted in both space and time.

References

Appadurai, A. (1996), *Modernity at Large: Cultural Dimensions of Globalization* (Minneapolis: University of Minnesota Press).

Basch, L. et al. (1994), *Nations Unbound: Transnational Projects, Postcolonial Predicaments and Deterritorialized Nation-States* (Amsterdam: Gordon and Breach).

Bauman, Z. (1998), *Globalization: The Human Consequences* (Cambridge: Polity Press).

—— (2000), *Liquid Modernity* (Cambridge: Polity Press).

Castells, M. (1996), *The Rise of the Network Society* (Berkeley: University of California).

Directorate of Labour (1999), 'Atvinnuleyfi 1998' (published online 10 Feb 1999) http://www.vinnumalastofnun.is (home page), accessed 15 March 2007.

Escobar, A. (2001), 'Culture Sits in Places: Reflections on Globalism and Subaltern Strategies of Localization', *Political Geography* 20, 139–74.

Faist, T. (2000), 'Transnationalization in International Migration: Implications for the Study of Citizenship and Culture', *Ethnic and Racial Studies* 23:2, 189–222.

Guarnizo, L.E. (2003), 'The Economics of Transnational Living', *The International Migration Review*, 37:3, 666–99.

Gupta, A. and Ferguson, J. (eds) (1997a), *Anthropological Locations: Boundaries and Grounds of a Field Science* (Berkeley: University of California Press).

—— (eds) (1997b), *Culture, Power, Place: Exploration in Critical Anthropology* (Durham and London: Duke University Press).

Hannerz, U. (1996), *Transnational Connections: Culture, People, Places* (London: Routledge).

Inda, J.X. (2000), 'A Flexible World: Capitalism, Citizenship, and Postnational Zones', *PoLAR: Political and Legal Anthropology Review* 23:1, 86–102.

Inda, J.X. and Rosaldo, R. (2002), 'Introduction: A World in Motion', in J.X. Inda and R. Rosaldo (eds).

—— (eds) (2002), *The Anthropology of Globalization: A Reader* (Malden: Blackwell Publishing).

Kearny, M. (1995), 'The Local and the Global: The Anthropology of Globalization and Transnationalism', *Annual Review of Anthropology* 24, 547–655.

Levitt, P. and Glick Schiller, N. (2004), 'Conceptualizing Simultaneity: A Transnational Social Field Perspective on Society', *International Migration Review* 38:3, 1002–39.

Massey, D. (2004), 'Geographies of Responsibility', *Geografiska Annaler* 86 B:1, 5–18.

Olwig, K. (2003), 'Transnational Socio-Cultural Systems and Ethnographic Research: Views From an Extended Field Site', *The International Migration Review* 37:3, 787–811.

Olwig, K. and Sørensen, N. (eds) (2002), *Work and Migration* (London: Routledge).

Parreñas, R. (2001), *Servants of Globalization: Women, Migration, and Domestic Work* (Stanford: Stanford University Press).

Sassen, S. (1991), *The Global City: New York, London, Tokyo* (Princeton: Princeton University).

Savage, M. et al. (2005), *Globalization and Belonging* (London: Sage Publications).

Skaptadóttir, U. (2004), 'Responses to Global Transformations: Gender and Ethnicity in Resource-based Localities in Iceland', *Polar Record* 40:214, 261–7.

Skaptadóttir, U. and Wojtynska, A. (2006), 'Gendered Migration from Poland to Iceland, Women's Experiences', Paper presented to the *6th European Gender Research Conference: Gender, Citizenship in a Multicultural Context* (Łódź).

Sørensen, N. and Olwig, K. (2002), 'Mobile Livelihoods. Making a Living in the World', in K. Olwig and N. Sørensen (eds).

Statistics Iceland (2004), 'Migration 1986–2003' Statistical Series, (published online 10 Feb 2004) http://www.statice.is/ (home page), accessed 15 March 2007.

—— (2007), 'Population: Citizenship and Country of Birth', statistical page http://www.statice.is/ (home page), accessed 15 March 2007.

Urry, J. (2001), *Sociology beyond Societies* (London and New York: Routledge).

Vertovec, S. (1999), 'Conceiving and Research Transnationalism', *Ethnic and Racial Studies* 22, 447–62.

—— (2004), 'Migrant Transnationalism and Modes of Transformation', *International Migration Review* 38:3, 970–1001.

Wimmer, A. and Glick-Schiller, N. (2003), 'Methodological Nationalism, the Social Sciences and the Study of Migration. An Essay in Historical Epistemology', *International Migration Review* 37:3, 576–610.

Chapter 11

Transnational Marriages: Politics and Desire[1]

Anne Britt Flemmen and Ann Therese Lotherington

Introduction

Regular crossings of the Russian/Norwegian border became a possibility again as a result of a loosening of control during the early 1990s. The new mobilities took different forms. During the first period, the most visible change in local people's everyday lives on the Norwegian side of the border was Russians arriving on short-term tourist visas to engage in petty trade in local markets. Their products were mainly Russian handicraft, from tablecloths to glassware, but also alcohol and cigarettes. Due to the economic divide between the two countries in terms of standards of living, simply selling their legal quota of alcohol and cigarettes was enough to pay the traders' travel expenses to Norway. With the Russian economy going through a huge transition period at that time, the Norwegian media reported very bad living conditions for people just over the border from Norway. As a result there were private initiatives on the Norwegian side to collect clothes, food and money for 'our neighbours' in the East. A big media outcry also developed, relating to prostitution in small rural areas. Local Norwegian men, as well as the Russian mafia, were targeted as being responsible for creating the market for Russian women selling sex, and hence exploiting poor women struggling to survive (Stenvoll 2002; Leontieva and Sarsenov 2003; Flemmen 2007).

Alongside these border-crossing activities and media discourses, other long-term mobility strategies emerged. However, restrictions on the mobility from Russia to Norway, imposed by the nation-states, made only some of these forms of mobility available to Russian citizens. As citizens of a non-European Economic Area country, the options for migrating to Norway on a long-term basis were connected to education, which was time-restricted; to work, that is, obtaining a contract as a specialist required in Norway; or to marrying a Norwegian citizen. This chapter is basically concerned with marriage as a strategy for mobility.[2]

1   This chapter has been written as part of the research project 'Women Crossing Borders', financed by The Research Council of Norway.

2   The analysis is based on qualitative material, consisting of around 30 interviews with Russian women and Norwegian men, in addition to data and analysis of family immigration and marriage patterns from Statistics Norway (for example, Daugstad 2006).

In other articles we have discussed different aspects of Russian/Norwegian marriages, such as the ability of Russian women to exert their citizenship rights (Lotherington and Fjørtoft 2006; Lotherington and Fjørtoft 2007); Norwegian men's constructions of gender and nationality (Flemmen 2004); the subject positions made available to Russian women and to Norwegian men related to them in Northern Norwegian newspapers, and how this constructs the Norwegian majority (Flemmen 2007); the Norwegian majority population's reaction to Russian/Norwegian marriages; the conceptual and analytical challenges posed in studying these marriages (Flemmen forthcoming), and finally the challenges Russian women face due to Norwegian immigration policy (Lotherington and Flemmen 2007). In the present work we are more concerned with analysing the marriage patterns between Norway and Russia at a macro level, as well as at the level of everyday life. We are particularly interested in exemplifying some of the complexities and the multi-directedness of the flows across the Russian/Norwegian border within the frame of transnational marriages.

Transnational Marriages

In general, transnational marriages have had a steady increase over the past 15 years (Constable ed. 2005). In Norway 900 such marriages were entered into in 1991, whilst in 2004 the number had increased to 3,000. Even so, foreign-born spouses count for only 5 per cent of Norwegian-born men and 3 per cent of Norwegian-born women's choice of marriage partner in 2005. The vast majority of both male and female spouses are Danes and Swedes, but beyond this a very different pattern arises regarding the choice of spouse. Norwegian-born men most often marry women from Thailand (3,589), the Philippines (2,873) and Russia (2,240), whilst Norwegian-born women marry men from the United Kingdom (2,750), Germany (1,522) and the USA (1,468).[3] Looking at the number of Russian/Norwegian marriages as of January 2005, there were 2,240 couples consisting of a Norwegian-born man and a Russian-born woman in Norway, and only 65 couples with a Norwegian-born woman and a Russian-born man (Statistics Norway, Daugstad 2006, 84).

Hence it is clear that the Norwegian marriage patterns are gendered in a certain way: men marry non-Western women and women marry Western men. The increase in cross-border marriages also basically relates to male non-Western marriages, that is, marriages with women from Thailand, the Philippines, Russia and Poland. The numbers for Russian/Norwegian marriages rose from zero in 1990 via 24 in 1995 to 323 in 2004 (Statistics Norway, Daugstad 2006, 96). They now seem to have stabilized at an annual level of 300–400 such marriages in Norway. From this it may be concluded that there has been a dramatic change in the marriage pattern between Norwegians without an immigrant background and Russian citizens, from the beginning of the 1990s to date, and that the increased number of transnational marriages has been mainly due to marriages between Norwegian-born males and Russian-born women.

---

3   All numbers refer to existing marriages.

When 'Western' men have married women from 'Eastern' countries, there has been a tendency in international literature to view these as 'mail-order marriages'. The suspicion is that wealthy men from the privileged West are looking for women from poorer countries in order to take advantage of them using their own power position as men, being relatively-speaking wealthier, and as citizens possessing language skills, as well as a knowledge of the rules and regulations of the country in which they reside. The assumption is one of misuse of a position of power on the part of the men. The women are often perceived to be lacking other possibilities and using marriage as a strategy to escape from poverty in their home country.

In her work on correspondence marriages, Constable (2003) felt the need to reconceptualize so-called 'mail-order' marriages and address the lesser-known but more common side of these marriages, rather than the heart-breaking stories related by the media. Her story, like ours, is one of less intrigue, less sex and violence and more everyday experiences relating to marriage migration.[4] Inspired by her conceptual framework, we have taken globalization and transnationalism as our conceptual starting-points, rather than more traditional and unidirectional approaches to migration. The concepts of transnationalism and transnational marriage are used to avoid assumptions of one-directional movement and a one-dimensional conception of power. The literature on transnationalism '*considers multidirectional flows of desires, people, ideas, and objects across, between, and beyond national boundaries*'(Constable 2003, 215–16). In addition to the important inclusion of flows of desire, we shall show that the emphasis on the multi-directional nature of the flow is important to a better understanding of existing marriages across the national borders between Northern Norway and North-West Russia, at a micro level as well as at a macro level.

Political Economy and the Cultural Logic of Desire

From the presentation of transnational marriages in Norway, two important macro trends are in need of explanation: the first is at *the level of the nation state*, while the second is at *the level of gender relations*. On the one hand this is a question of increased intermarriage between Russians and Norwegians and, on the other hand, of the differences in transnational marriage patterns between Norwegian men and Norwegian women: that is, the one-sidedness of the gender exchange between Russia and Norway.

In order to understand the increase in the number of intermarriages between Norwegian and Russian citizens at a macro level during the second half of the twentieth century, it is useful to analyse them as the result of globalization processes and changes at the level of political economy. Recognizing the gendered geographies of power, and what Massey calls the 'power geometry' at work on a global scale, we are concerned not only with who moves but also how people are differently located when it comes to access to and power over movement (Massey 1994, 149).

---

4   This is not to deny that violence and trafficking exist (see Brunovskis and Tyldum 2004, Bjerkan ed. 2005). Assumptions about Norwegian men's widespread abuse of Russian women through serial marriages have not been substantiated, however (Lidén 2005).

The Russian/Norwegian couples we have studied possess the 'network capital' (see Chapter 8 by Larsen and Urry) necessary to enable such movement. By network capital they refer to people's *'facility* for "self-directed" corporeal movement and communication at-a-distance'. This refers to the capacity, both technical and social, to enter into and sustain social relations with people who are not necessarily proximate.

It is, however, important to take into account the way in which the political economy intersects with everyday lives. A focus on the political economy illuminates the circumstances that made this particular mobility possible (the dissolution of the Soviet Union), the political relationship between the former Soviet Union and the Western world at present (affecting immigration regulations) and the current economic situation in Russia as well as in Norway (restricting free mobility between Russia and Norway). However, we also need to explore how these circumstances are handled in people's everyday lives. We shall return to this in a moment.

In order to understand the difference in marriage patterns between Norwegian-born men and Norwegian-born women, the one-sidedness of the gender exchange, it is useful to analyse them partly as the result of what Constable (2003, ed. 2005) calls a cultural logic of desire. Discourses and stereotypes about foreign women, in particular non-Western women, often portray them as more traditional and family-oriented than Western women on the one hand, whilst on the other hand connecting them to prostitution, sex work and trafficking (Stenvoll 2002, Leontieva and Sarsenova 2003, Flemmen 2007). Discourses about 'Western' men in non-Western countries might construct these men as being more modern and open-minded when it comes to gender relations, in contrast to the local men's more traditional and patriarchal gender expectations (Flemmen 2004, Nordin 2007). Hence we see a desire for traditional wives on the part of Western men and for modern husbands on the part of non-Western women. This has led Constable (ed. 2005, 7) to suggest that the current global 'marriagescape' both reflects and is propelled by fantasies about gender, sexuality, tradition and modernity. As we have discussed elsewhere (Flemmen 2004), some of our informants have stated similar expectations. Although these desires might be viewed as part of the power geometry between Russia and Norway, seen in strictly macro-economic terms, we are concerned with studying the couples as a diverse group, with different opinions, expectations and motivations. A focus on the couples' everyday practices is therefore an important supplement (Bryceson and Vuorela 2002). The couples must be understood in the particular historical and global context in which they appear, and as people who both exert power and are subject to it (Constable 2003).

We shall suggest that some of the Russian women's understandings of Western or Norwegian men, and of the imagined West they represent, are part of the cultural notion of desire that is made thinkable, desirable and practicable by a wider political economy (Constable 2003, 120).[5] What is distinctive about Russian/Norwegian relationships is not however, that they involve pragmatic and practical concerns (all marriages do) (see Thagaard 2005 for a Norwegian example), but rather that they

---

5   The same, of course, is true of some of the Norwegian men's understanding of Russian women.

allow the contradictions and paradoxes to become apparent. Where love is simply taken for granted in Norwegian marriages, and most often in marriages between people from Western countries, love is assumed to be absent in the Western/non-Western marriages. Constable (2003, 120) suggests that marriages such as Russian/Norwegian ones thus threaten to reveal tensions that other marriages more easily ignore or mystify.

Encounters at the Level of Everyday Life

While migration during the seventeenth century was said to contribute to the development of national movements in the countries the migrants moved *to*, resulting in the migrants' personal and emotional identification with public ideology in these countries, migration today is more often described as 'long-distance nationalism' or as a way of maintaining identities attached to the countries that people have moved *from*. The consequences of migration are therefore assumed to be increased local heterogeneity and a tendency to sustain permanent social and cultural connections across national borders. At the same time, the general globalization process is presumed to result in cultural and personal identity becoming less attached to the nation-state (Fuglerud 2001, 193–5). The more specific question directed at the level of everyday practice is what multidirectional flows of desires, ideas and people are being shaped within Russian/Norwegian marriages in Northern Norway at the beginning of the twenty-first century?

The forms of mobility made available to Russian women are limited by national and international regulations. The options for migrating to Norway on a long-term basis are, as previously mentioned, connected to education (time-restricted), work (contract as a specialist needed in Norway) or marriage (marrying a Norwegian citizen). In our material we find all these different forms of migration. Of the Russian women we have interviewed, some originally moved in order to obtain an education or to work as a specialist, and only later found a husband, while some moved as a direct result of marrying a Norwegian or Western man living in Norway. One couple met as a result of the Norwegian man's labour migration to the former Soviet Union, and decided to move to Norway later.

Two places to call home

Living with his mother until her death, Kjetil was conscious of the emptiness of the house and felt lonely. Due to health problems, he was no longer able to keep up his work. He met Svetlana in Nikel, a town just one hour's drive across the Russian border from Norway. Svetlana was born in Siberia and educated in St. Petersburg, and had moved to Nikel in order to find work.

Being a Sámi with Russian ancestors, Kjetil feels a connection with Svetlana and the Russian people. To him, his marriage represents a way of creating a link with his own past. Kjetil holds Svetlana in high regard and is very proud of her; she is clever with people, as well as with her work. Svetlana and Kjetil have a house in Norway and an apartment across the border, where they spend quite long periods when she has time off work. One of the neighbours looks after the apartment and pays the different bills (electricity, telephone, and so on) when they are not there. Svetlana holds dual citizenship and she follows the political debates and votes in both countries. The couple appreciate being able to spend time in Russia. Kjetil is very fond of Russian food as well as the people, and he is particularly happy about the close personal relationships with family and friends, as well as the emotional closeness between people that he experiences there.

An additional benefit of living so close to the border is that they are able to go to the doctor and the dentist in Russia since, in their view, the Norwegian health care system is very poor. The long waiting lists and the reluctance of Norwegian doctors to prescribe patients the pills and check-ups they want and need amount to a scandal, in their opinion.

With her dual citizenship, Svetlana feels that she acts as a kind of cultural translator. In Norway she explains what it is important to know about Russia, and in Russia she tells people what it is important to know about Norway. The dual citizenship, in her opinion, makes it easier to achieve co-operation between the countries.

This family sustains a much closer connection with Russian society than most of the couples we have talked to. This may have something to do with the fact that they do not have children. Neither of them has many close family members alive, though Svetlana appreciates the contact they have with his relatives in their everyday life in Norway. However, the couple does have a social network in Russia with whom they keep in regular touch.

To Svetlana, her connectedness to Norway is achieved through her dual citizenship,[6] as well as the position she has developed for herself as a cultural translator between the two countries. The sense of connectedness to Russia felt by Kjetil relates partly to sections of his family being of Russian Sámi origin and is partly a result of their frequent stays in Russia.

The migrant connection to Russia is varying in degree. Some of the migrants try to visit their parents as often as possible and familiarize their children with Russian society. Others make sure that their visits to Russia involve cultural experiences, such as concerts, exhibitions and operas, which are what they miss living in a part of Norway with few urban facilities. The last group, amongst whom we find women without strong family ties in Russia, very rarely travels to Russia at all.

Even though the women maintain close connections with Russia, either through visits, email contact or phone calls, they seem to have decided that they are in Norway to stay. Those who have children find it difficult to imagine leaving their children in Norway if they themselves should decide to move back to Russia. Recognizing that their children are Norwegian, that they would not be able to keep up with the others or adjust to the Russian school system after attending the Norwegian school system for a number of years, and considering them to be ill-equipped for life in Russia generally, the women foresee a future of growing old in Norway. One Russian woman told us that even though they did not have children, she and her husband had decided to be buried next to each other in Norway. The couples find life in Russia too complicated, even though some of them have had serious discussions about moving there. Some have also discussed moving to a more urban area in Norway. So, as we

---

6   The Norwegian regulations concerning dual citizenship are currently in the process of change.

can see from this small material, there are different ways of sustaining social and cultural connections across the national border. With new Norwegian regulations concerning the possibility of holding dual citizenship, and with improved living conditions in Russia, it is reasonable to believe that new negotiations between forms of movement and belonging will be necessary for these couples.

Network communication

It is considered distasteful to connect politics and market forces with personal lives and intimate relationships in Norwegian and many other Western societies (Zelizer 2005). This aversion may be linked to a common domestic-public split in Western countries, where the family is a refuge from impersonal political and economical forces (Constable 2003, 116). The Russian/Norwegian marriages must, however, be understood in the light of the dissolution of the Soviet Union, as well as the collapse of the Soviet welfare system. Life in post-Soviet society has been very difficult. Periods of extreme shortages of goods such as food and clothing, as well as lack of money, both because of unemployment and because workers were not paid, created a border area between Norway and Russia that represented one of the deepest welfare gaps in the world.[7] For some of the Russian women, such as the woman referred to below, marriage migration has to be understood as an attempt to improve their own welfare as well as their children's opportunities.

> Maria and Roger first met after communicating at a distance. Maria placed an advertisement with a Russian contact agency and Roger's Russian neighbour Anastasia was contacted by the marriage agency manager, who was her friend. Anastasia was, after all, already married to a Norwegian man and living in a small community in Norway, so the manager looked to her for help in finding a Norwegian man for Maria. Knowing he was single, Anastasia approached her neighbour Roger to enquire how he felt about getting in touch with a Russian partner. Since he had already been thinking about looking for a Russian woman, Roger agreed to accept Maria's phone number. After some time he decided to call her, but after trying to communicate in a mixture of English, Russian and Norwegian, using dictionaries, they asked the neighbour Anastasia to be a mediator and interpreter as they continued to communicate by letter in their own languages. Upon visiting Northern Norway for the first time, an anxious and excited Maria decided that if Roger brought her flowers when he came to pick her up from the airport, he had to be a good man. He brought her flowers and they married during her second trip to Norway.

> After her first husband had died Maria had held down two jobs, but was still unable to foresee a future where she could support her daughter's university education in North-West Russia at the end of the 1990s. In order to secure better prospects for her teenage daughter, Maria looked for a foreign man. When we met them, they had been married for two years. After the first year, when Maria had served as crew on Roger's small fishing boat, she spoke almost fluent Norwegian and is at present at home taking care of their small daughter, teaching her the Norwegian language. Doctors have told her that she may teach her daughter Russian at the same time, without this being confusing for the child. However, Maria wants to wait until her daughter starts school before teaching her Russian.

---

7 However, these economic differences were more acute during the 1990s than they are at present.

She and Roger both took great pride in her teenage daughter being awarded the best grade in her class in Norwegian language, and this after only two years in the country! As far as her own future is concerned, Maria dreams of starting her own business, importing items from Russia and selling them in Norway. Roger supports the idea and has volunteered to be the driver between Russia and Norway.

Maria and Roger met as a result of network communication at a distance. Being very matter-of-fact about how they met, Maria explains that she was unable to give her daughter the life she wanted her to have in Russia. Maria's motivation for finding a Western man was the result of differences in opportunities represented by a life in the West. For Roger it was more a question of finding a companion in life and starting a family. We did not specifically ask Roger why he was looking for a Russian wife, but since Russian women had married into his community this may have influenced his desire to do so. Maria and Roger seem to have been able to create a good life for themselves and their two daughters. Maria's sixteen-year-old daughter has had the opportunity to get the education that her mother wanted so badly for her, while Maria at the time of the interview was at home taking care of the youngest daughter. However, Roger thought that she ought to attend a kindergarten soon, in order to learn to adjust to other children. Entering the labour market one year after giving birth seems too soon for Maria, who is used to women having three years' maternity leave. Spouses having separate wallets and economies is another thing that seems strange to Maria. In her experience from Russia, the husband brought the money to the wife, who would manage the family finances. The gendered division of labour and the ways of being women and men are different in Norway and Russia (see Flemmen 2004).

When Maria contacted the marriage agency in Russia she was looking for a Western man. The fact that he turned out to be a Norwegian was more or less accidental. We do not, however, interpret her type of mobility as the result of a general fascination with the West. Roger, on the other hand, had specifically considered looking for a Russian wife. We understood his main concern to be securing a partner and family. Other men we have interviewed, on the other hand, have specifically looked for Russian wives, whom they believed to be more family-oriented than Norwegian women. Some men felt that finding a Russian woman would improve the chances of the marriage lasting (Flemmen 2004). For other men, however, there seemed to be other issues at stake altogether.

Complex flows of desire

One of the unexpected results of our research relates to the complexities of the flow of desire, people and ideas across and between national borders. Whilst our attention was initially focused on Russian women as being central to this movement, the couples' stories revealed a more complex picture. Even though Russian women had initiated the movement in many instances, Norwegian men had done so as well. Some men had met their wives on short, work-related trips to Russia, whilst one man had met his wife in the former Soviet Union. Two of the Norwegian men turned out to have long personal histories of fascination with the former Soviet Union. For one of these men, his political interest caused him to move to the former Soviet Union before its dissolution. Still struggling with the language, he met his wife at his workplace and they fell in love. They married, had children and lived in the former Soviet Union for several years before moving to Norway in the late 1990s, partly as a result of the difficult times there. For the other man, Tom, the fascination of the Soviet Union had been shared by three generations of his family, but he himself had not spent much time in the former Soviet Union at the time when he met Larissa.

> Growing up in a family with long-standing communist sympathies and a close relationship with the former Soviet Union, it is perhaps not surprising that Tom's second wife is Russian. Larissa had been in Norway for a number of years and was working in a hotel when she and Tom met at a party hosted by acquaintances. Since the age difference between them was about twenty years, her parents were very sceptical when she announced to them that she wanted to marry him. After meeting him, however, they did accept him and now support the union of their daughter with a man of their own age. He finds that her way of being a woman makes it easy for him to be a man, quite unlike his former experiences with Norwegian women. Having been brought up in Russia, a society with more distinct gender roles, she has, in his view, a clearer sense of the differences between women and men.

> Even though he himself had not been able to go to youth camp in the former Soviet Union, as his older brothers had done, he has a deep fascination and very knowledgeable relationship with Russia. In fact, he believes that what Larissa found attractive and different about him was his knowledge of Russian literature and the country's history. One of the things they enjoy doing together is travelling to Moscow and St. Petersburg to go to the ballets and to concerts, museums and art galleries. Tom's knowledge and appreciation of Russian culture is something he talks about with great enthusiasm. The Russians know how to appreciate the best that culture has to offer and maintain very high standards in education, as well as in the arts. Living in a part of Northern Norway that he describes as a cultural desert, where 'the only living idea is fish', they have discussed moving to Russia but have decided against it. Life in Russia is too difficult and unpredictable. Tom and Larissa have both lived in several places in Norway, but for the most part they have been located in the northern part of the country.

Tom's connection to Russia is still strong, particularly at the level of its culture and mentality. He does, however, emphasize the general cultural aspect of the connectedness more than maintaining a social network with specific people.

The flow of desire between Russian women and Norwegian men can thus take several forms. It is the result of complex constructions of gender and nationality. For some, as in Tom's case, the fascination at the level of the nation preceded the fascination of a specific woman/man; for others, the desire for a specific woman/man was closely related to nationality; for still others, the desire for the woman seemed to be unrelated to her nationality. Some of our male informants did, however, express concern about Norwegian men looking for specifically Russian women. They feared that the men intended to exploit the women. In the case of some of the couples interviewed, the expectations regarding gender relations were constructed rather differently. One Norwegian man explained that his wife, as a feminist, needed to find a Norwegian man, since no Russian man would have tolerated her. Since Norwegian men were expected to be more 'modern' and used to gender equality, she stood a better chance of finding a suitable man in Norway (Flemmen 2004). A similar logic was also applied by a Norwegian man to explain why divorced or widowed Russian women over 35 years old found it easier to find a Norwegian man to marry than a Russian. In both instances, the Norwegian man is presented as the solution to the Russian woman's difficulty in finding a Russian partner at home. Their desires were diverse: we have seen Norwegian men longing for simply a woman, for a Russian woman in particular, or for a family-oriented woman, and we have seen Russian women desire a man in general, a Western man, a more equality-oriented man, and a better life for themselves and their children.

Concluding Remarks

In this chapter we have analysed the transnational marriages between Russian women and Norwegian men on two levels. We have analysed them at the level of political economy and cultural logic of desire, in order to understand the changes in Norwegian/Russian marriage patterns at the level of the nation-states and the one-sidedness of the gender exchange. In order to incorporate the logic of the migrants we have included examples of the couples' everyday practices. Through these examples, and others referred to in other parts of our work, we hope that the power relations made visible at the macro level have been rendered more complex, and the migration strategies more human.

Constable (ed. 2005) suggests that we are currently seeing a global marriagescape being shaped, limited by existing economical, political and historical factors. The majority of the marriage migrants from Russia to Norway are women, as are most of the international marriage migrants. Marriage migration is thus gendered. Most of the international marriage migrants move from poorer to wealthier countries. This suggests that marriage migration is shaped by political economy, but other factors such as 'a cultural logic of desire' may also be important (Constable ed. 2005). Western men's assumptions about and desire for traditional non-Western women, as well as non-Western women's assumptions about and desire for modern Western men, form part of this picture. However, we want to stress that pragmatic and practical considerations in cross-border marriages do not necessarily preclude love and other non-material forms of desire.

References

Bjerkan, L. (ed.) (2005), *A Life of One's Own. Rehabilitation of Victims of Trafficking for Sexual Exploitation* (Oslo: Fafo-report 477).

Brunovskis, A. and Tyldum, G. (2004), *Crossing Borders. An Empirical Study of Transnational Prostitution and Trafficking in Human Beings* (Oslo: Fafo-report 426).

Bryceson, D. and Vuorela, U. (eds) (2002), *The Transnational Family. New European Frontiers and Global Networks* (Oxford: Berg).

Constable, N. (2003), *Romance on a Global Stage. Pen Pals, Virtual Ethnography, and 'Mail-Order' Marriages* (Berkeley: University of California Press).

—— (ed.) (2005), *Cross-Border Marriages: Gender and Mobility in Transnational Asia* (Philadelphia: University of Pennsylvania Press).

—— (2005), 'Introduction: Cross-Border Marriages, Gendered Mobility, and Global Hypergamy', in Constable (ed.).

Daugstad, G. (2006), *Grenseløs kjærlighet? Familieinnvandring og ekteskapsmønstre i det flerkulturelle Norge.* Rapport 2006/39 (Oslo-Kongsvinger: Statistisk Sentralbyrå).

Flemmen, A.B. (2004), '"For meg er det lett å være mannfolk i lag med Elena". Norske menns konstruksjoner av kjønn i russisk-norske ekteskap', DIN. *Tidsskrift for religion og kultur*, no. 2–3: 64–73

—— (2007), 'Russiske kvinner i nordnorske aviser – minoritets og majoritetskonstruksjoner', in *Tidsskrift for kjønnsforskning*, no. 1: 37–54.

—— (forthcoming), 'Transnational Marriages – Empirical Complexities and Conceptual Challenges. An Exploration of Intersectionality', *NORA – Nordic Journal of Feminist and Gender Research*, 16:2.

Fuglerud, Ø. (2001), *Migrasjonsforståelse. Flytteprosesser, rasisme og globalisering* (Oslo: Universitetsforlaget).

Hvinden, B. and Johansson, H. (eds) (2007), *Citizenship in Nordic Welfare States: Dynamics of Choice, Duties and Participation in a Changing Europe* (London: Routledge).

Leontieva, A. and Sarsenov, K. (2003), 'Russiske kvinner i skandinaviske medier', *Kvinneforskning*, no. 2: 17–31.

Lidén, H. (2005), *Transnasjonale serieekteskap. Art, omfang og kompleksitet.* Report 2005:11 (Oslo: Institutt for Samfunnsforskning).

Lotherington, A.T. and Fjørtoft, K. (2006), 'Russiske innvandrerkvinner i ekteskapets lenker', *NIKK-magasin*, no. 1: 30–31.

—— (2007), 'Gender, Capabilities and Democratic Citizenship', in Hvinden and Johansson (eds).

Lotherington, A.T. and Flemmen, A.B. (2007), 'Ekteskapsmigrasjon i det norske maktfeltet', *Sosiologi i dag*, 37: 3–4, 59–82.

Massey, D. (1994), *Space, Place, and Gender* (Cambridge: Polity Press).

Nordin, L. (2007), *Man ska ju vara två. Män och kärlekslängtan i norrländsk glesbygd* (Stockholm: Natur & Kultur).

Stenvoll, D. (2002), 'From Russia with Love?: Newspaper Coverage of Cross-Border Prostitution in Northern Norway 1990–2001', *The European Journal of Women's Studies* 9:2: 143–62.

Thagaard, T. (2005), *Følelser og fornuft. Kjærlighetens sosiologi* (Oslo: Abstract forlag).

Zelizer, V.A. (2005), *The Purchase of Intimacy* (Princeton: Princeton University Press).

Chapter 12

The Svalbard Transit Scene

Arvid Viken

Introduction

This chapter is about extreme mobilities – in Longyearbyen on Svalbard, one of the most northern and remote inhabited locations. Svalbard was not meant for human beings: it is a human conquest that was made back in the early sixteenth century. After its 'discovery' (in 1596) the Dutch and the English started to hunt the mammals in the ocean around to provide Europe with ivory (from the walrus) and oil (from the whale). This exploitation resulted in the disappearance of several species of whale from the area. Today, this Arctic region is facing a similar threat: the oil industry has a strong focus on the northern periphery as an area for exploration and possible future exploitation. In the interim, Svalbard has been an area for trapping seals, polar bears and polar foxes, a mining industry site and an arena for research activities and tourism.

Longyearbyen is the hub of all Norwegian activity on Svalbard, encompassing the District Governor's Office, a university campus, a research centre, a significant tourist industry, a viable service sector and a coal mine (see Figures 12.1–12.3). It is also a place of extreme mobility: miners commuting to Svea, 50 kilometres away; many public employees contracted for only two years; guest researchers staying for short periods; students coming for only one or two semesters; and in tourist terms, many of the employees and all of the customers staying for short periods. Yet the most significant sign of mobility is the turnover in the population: 20 per cent in 2005, 25 per cent in 2006 (Longyearbyen Community Council 2006), compared to 2.4 per cent – measured as moves between counties – on the Norwegian mainland (Statistics Norway 2007). Thus, Longyearbyen is obviously a place where mobility is the norm and fixity the exception (cf. Cresswell 2006). It is also a partly global place: the researchers and the students, but also people in other sectors, come from all over the world, and so do the tourists. Present-day Longyearbyen is far from being the isolated place it was before 1975, when the airport opened. With its modern, air-based accessibility, contemporary Svalbard is a good example of what is enabled by post-modern time-space compression: a modern and globally integrated life on the edge of the earth. Longyearbyen is both influenced by and adapted to contemporary tendencies such as mobility and globalization, and merits labels such as 'fluid', 'liquid', 'mobile', 'touring', 'network society' and the like (Bauman 2001; Castells 2004; Clifford 1997; Urry 2000; Wellman and Berkowitz 1988; Wellman 1999).

This chapter focuses on the fact that Longyearbyen is a place of mobilities, a place for fluctuating relations and changing identities, and a place where the turnover in the population is far higher than the average in Norway. Firstly,

Svalbard is presented with regard to its location and a brief overview of the past is provided. Then a theoretical background is presented, covering social network and social capital perspectives. The empirical section covers three aspects of living in a mobile community: what characterizes living in such a mobile place, the way into the community and identification with the island. The data used for these analyses comprises three focus group interviews and six in-depth interviews, undertaken in March 2007. The focus groups were put together according to two criteria: time spent on Svalbard (two groups) and area of work. There are of course a series of methodological strengths and weaknesses relating to this method (cf. Morgan 1996), and one is concerned with a lack of representativity. However, it is maintained that when many express the same view, there are reasons to believe that this reflects a general tendency (Macnaghten and Jacobs 1997).

Figure 12.1   Longyearbyen: a residential area

Source: photo by Ole Magnus Rapp.

Svalbard: A Community on the Northern Rim

Svalbard is located in the Arctic Ocean. It is one of the northernmost inhabited places on earth, stretching from the 74th to the 81st degree of latitude. The archipelago covers a land area of 61,229 square kilometres, with a coastline that is over 3,500 kilometres long. The climate is cold but, thanks to the Gulf Stream, not as cold as many other locations in the north. The average air temperature is +6° C in summer and –14° C in winter. The sea temperature hovers between –1° C and +4° C. In the winter the sea is normally frozen.

Figure 12.2   UNIS: the university centre

Source: photo by Ole Magnus Rapp.

Figure 12.3   An excellent area for recreation and tourism

Source: photo by Ole Magnus Rapp.

Svalbard was 'discovered' by the Dutchman Willem Barents in 1596 (for more historical detail, see Arlov 1996). The industrial history of Svalbard has five distinguishable eras that partially represent separate activities. The first period, from 1603, was initially dominated by Englishmen hunting walrus, followed by whale hunting – primarily to extract the oil – by the Dutch from 1611, but later hunters from England, Spain, France, Germany and Denmark (including Norway) took part in this industry as well. The second period stretches from 1700–1850, a period dominated by Russians trapping mammals primarily for the sake of their fur. The third period extends from the beginning of the nineteenth century, when Norwegians more or less took over the trapping activities. This was also a period of exploration of the north. The fourth period started around the beginning of the twentieth century, an epoch encompassing the search for and excavation of minerals, with coal-mining as the only commercial success. The fifth period started towards the end of the 1980s and relates mostly to the position and development of Longyearbyen, the main Norwegian settlement. This had been a mining town since the early twentieth century, but towards 1990 the decision was taken that coal-mining should be scaled down and replaced by other industrial activities, with research, education and tourism to be given particular priority. Along with changes to its industrial platform, Longyearbyen has undergone an extensive process of modernization and 'normalization', including societal systems supporting family life. As a 'normal' environment the town has become more complex, with a welfare system, infrastructure and different types of regulations, including restrictions on terrestrial travel to reduce the pressure on nature, partly due to a growing level of tourism (Viken 2006). This aspect of normalization has therefore not been universally praised.

Communities, Social Networks and Social Capital

In studying what it is like to live in a mobile community, several theoretical perspectives may be chosen. One of them focuses on networks and ties, mechanisms vital to people's careers and wellbeing. Two theoretical approaches that deal particularly with such questions are social network theory and the theory of social capital. One individual who has contributed to the study of social networks is Barry Wellman (1988, 1999, 2001). 'I define 'community' as networks of interpersonal ties that provide sociability, supports, information, a sense of belonging and social identity', he states (2001, 228). One of his claims is that sociologists have been focusing on communities and relating the term to neighbourhoods, thereby ignoring the significance of long-distance contacts and mobility. He claims that by focusing on networks there are no territorial boundaries, however '[t]his does not mean that community has been lost but that it is much less likely now to be locally based and locally observed' (Wellman 1999, 19). Urry (2000, quoting Hetherington 1997) goes even further, claiming that local communities should be defined in relation to different types of mobilities: they move around within networks of agents, human and non-human. Urry and his colleagues also look at travel as a major aspect of the modern network society (Larsen et al. 2007).

In one of his accounts, Wellman (2001) shows how communities and networks have changed from being based on door-to-door and place-to-place contacts to becoming person-to-person relations. This implies that contemporary networks have been liberated from place and become a personal matter, a tendency enforced by personalized technology such as mobile phones and email, according to Wellman. He also points to another tendency, that many networks exist not between individuals but between roles. This also implies that networks are often narrow and specialized (Wellman 1999, 23; 2001, 245). Wellman claims that roles make it possible to interact without involving other people, and that this is made easier by the world of emails. One might also maintain the opposite – that acquiring a role may be a step into a network which can become vital for personal matters – and this is partly the focus of social capital theory (Putnam 2000).

Another implication in the modern network society is that 'private intimacy has replaced public sociability' and 'communities have become domesticated and feminised' – the last point referring to men's stronger involvement in domestic matters and women's more vital roles in societal affairs (Wellman 2001). These general assumptions imply that communities have been transformed into a series of networks, far less related to neighbourhoods than in earlier times. In Chapter 8 of this book, Larsen and Urry argue along the same lines, that social capital theory tends to over-focus on the neighbourhood aspect. They launch the term 'network capital' as an alternative, defined as 'the capacity to engender and sustain social relations with individuals who are not necessarily proximate, which generates emotional, financial and practical benefit'.

Wellman (1999, 18–19) claims that there are two major approaches within the study of social networks: as whole networks, or as personal communities. Social networks of the latter type are the most common and 'enable the researchers to study community ties, whoever with, wherever located, and however structured. They focus on the inherently *social* nature of community...' and study 'the social support that community networks provide: the supportive resources that community ties convey and their consequences for mental and physical wellbeing and longevity' (Wellman 1999, 19–20). Wellman distinguishes four spheres of personal networks: co-workers, friends, neighbours and clan members. In each of these spheres people may have intimate and nonintimate ties (Wellman 1999, 20). Others, however, use the term 'friend' as a category that indicates a degree of intimacy in a relationship. Ferrand et al. (1999) operate with a split between friends and confidants, the latter category representing the closest ties, whereas Savage et al. (2005) split between acquaintances and friends, friendship being most intimate.

Although these social capital theories are criticized (cf. Pawar 2006), there are some aspects of this theory that it is important to include in network analyses. Network membership can be a kind of individual (and collective) resource. Social capital analyses have flourished over the past two decades, very much due to the work of scholars like Bourdieu (1977, 1986), Coleman (1988) and Putnam (2000). Glover and Hemingway summarize Bourdieu's way of using the term as follows:

> Social capital lies in the persistent social ties that enable a group to constitute, maintain, and reproduce itself. Such ties...allow group members potential access to resources

held by others in the group...Bourdieu regarded social capital as purposive, a resource facilitating individual action by virtue of the individual's location within social networks and groups. (Glover and Hemingway 2005, 389)

Putnam (2000), who has received significant attention for his macro analyses of social capital, also perceives social networks as a resource for the individual. However, it is also maintained that '[a]ccess to resources in a social network is not automatic: knowledge of them and skill in navigating the network are required. [...]' (Hemingway 2006, 346). Thus, there are power and status relations involved in networks, and being tied to a network is not a guarantee of occupying a vital position.

Longyearbyen: An Open-minded Hierarchy of Time

Mobility is more or less institutionalized on Svalbard. Most people stay for only a year or two. Nature and outdoor recreation are the main motives for going there. Several informants claimed to be some sort of adventurer. It was said that if you were not fond of nature activities you should not go there, and you would be unlikely to stay. Another side of being an adventurer is being a traveller. Several informants maintained that many people on Svalbard are advanced tourists and globetrotters, travelling all over the world. Many of them go to other places on earth to practise a special interest. During the interviews, people told of others who regularly hunt bears in Alaska, paragliders who travel to Southern Spain to practise their sport, some who used to go to Las Vegas to play poker, and about people visiting golf courses all over the world. Travelling is also a way of keeping up social contacts with family and friends on the Norwegian mainland and ex-Svalbardians. So is hosting visitors. This can in fact be a bit too much: the exoticism of the islands attracts more than just close friends and relatives. 'If you wish, you can fill up your house all the time', several informants claimed.

Socially, people have different networks: most often, one consists of family and friends from other places where the informant used to live; a second comprises old Svalbardians and a third encompasses workmates and other contemporaneous inhabitants. 'Put it this way', one lady said, 'I am in four networks: one here on Svalbard, one relating to family, one network of women living in different places in Norway – companions who meet once in a while – and a job-based network.' This is probably relatively typical: different networks serve different purposes and offer different types of social support. The lady quoted above emphasized that her networks did not overlap. The pattern referred to may be a sign of weakening residential ties, and more dispersed and individually-based relations forming part of a liberated community (cf. Wellman 2001). The networking of Svalbard people, often including travel, adds to the mobility character of the town. There also seems to be a correlation between staying in a mobile community and living a mobile life.

With the high turnover that exists, Longyearbyen is a place for newcomers, where everyone is used to people coming and going. The majority of the inhabitants are in fact neophytes. Therefore people know what this situation is like. Newcomers are welcomed by many and are perceived as a resource: new faces in the pub, new members in the sociations, new friends and, as a manager of the local authority said, 'This is a dynamic community, new people constantly sharpen your mind and bring in new perspectives and knowledge.' However this depends on the individual life situation and phase: young people and those who are newcomers themselves are more open than the veterans. 'As a young person you are on the search, and much more open to new relations', it was maintained, as well as arriving during a period in life when enduring social networks are normally being knitted. One interesting aspect of social life in Longyearbyen is that social cohorts seem to exist: people tend to keep the friends they made during their earliest period on Svalbard. Thus, the initial period is particularly important. It was also said that there is a limit to how many friends one can absorb.

There is an obvious fatigue relating to the turnover: a limit concerning how long people can cope with getting to know new people and, worse, a limit to how often one is willing to take on the burden of mourning over vanished friends. All the veterans emphasized this. Leaving Svalbard has its own expression – *reise på slutt*, to leave and be done with it – which has a series of connotations. The leave-taking of people that have stayed there for years means a severe change in close relations and social life, and can be the end of a friendship. As has been stated generally, friendship is closely related to activities people undertake together (Giddens 1991); on Svalbard in sociations (cf. Urry 2000), on tour, at work. Thus, for many it is a situated friendship, but obviously not for all. The veterans particularly emphasized that they maintain contact with many ex-Svalbardians. As one said, 'I can travel all around Norway staying with old friends from Svalbard.'

People on Svalbard have few intimate friends (or confidants), and fewer over the years as people leave, but many acquaintances (non-intimate friends). Because it is a small place, the acquaintances seem to be closer than in most other places. In fact, friends and acquaintances are extremely important in Longyearbyen, as they substitute for family and kin in terms of social support and aid exchange. The normal pattern is not an option – aid exchanges are twice as frequent in kin-based webs as in friendship webs, according to a French study (Ferrand et al. 1999, 194). Having no relatives in town is not a problem: 'if necessary, I can leave my children in the custody of my friends', an informant said. So, even if it is recognized that these may be temporary friendships, people seem to be pretty close to one another.

One aspect of life on Svalbard that was emphasized by several informants is that you are treated as the person you are, not according to your genealogy. 'Where I come from, I am made responsible for the deeds of my ancestors over 400 years. Here, your family background is irrelevant', one informant claimed. This means that people do not encounter biased attitudes, as in many other places. 'Here, you are valued for what you do and stand for, not from where or whom you originate' was another comment. Social capital is not inherited, but created through people's own actions. However, you are supposed to conform to the local norms. You are not supposed to boast about what you did before coming to Svalbard. Another sign of local adaptation is that you wear expensive sportswear, even at work. More importantly, you should be able to tell stories from your experience in the wilderness that contains at least one of the following elements: polar bears, rough weather conditions, snowmobile, speed and driving problems, or hunting events. And the stories should refer to your own experience. Thus it was maintained that narratives are the ticket to social acceptance. This is probably more typical than exceptional. In fact, narratives expose identities, and on Svalbard this is a way of expressing a locally-based biography. 'In a particular socio-cultural environment, the self is given content, is delineated and embodied, primarily in narrative constructions or stories', according to Rapport and Dawson (1998, 28), quoting Kearney (1991, 3) and stating that 'it is in and through various forms of narrative emplotment that our lives...our very selves – attain meaning'. This is obviously also the case on Svalbard.

Another central aspect of the social life in Longyearbyen is that the workplace seems to be a node in most people's networks and friends are most often made among colleagues. Studies elsewhere show that work is a common base for different types of friendship, but that only accounted for about 40 per cent in the French study referred to above (Ferrand et al. 1999). In Longyearbyen this is obviously more significant, mostly due to the fact that the job is the overwhelming reason for moving to Svalbard. In addition, many stay for too short a time to transgress the work enclave. Without a job – or a partner with a job – it is not convenient to continue staying, amongst other things because housing is generally provided by the employers, or is extremely expensive. Thus, Longyearbyen is still a work-centred community and there is a tendency towards job-based social clusters. However, work also seems to be a gateway to other types of networks: to the many sociations. To some extent, these counteract the tendency to work-centred relations, but one informant said 'I feel that I am looked upon as deviant by my colleagues, I do not mingle with them. I am too occupied with the people in the choir.' For those coming to Svalbard as a partner and without a job, the sociations seem to fill an important social function. 'I'm so busy', a young, well-educated housewife claimed. 'Here I have so much free time...I am playing in a big band, do aerobics, play badminton, take part in a book-club and in a knitting-club, and meet people I know at *Fruen* ("The Missus" – a café)...So I am in fact rather busy...'

Longyearbyen used to be a community divided by class, with the management and administrative staff of the mining company ranked over the miners. The social differences were symbolized by accommodation being allocated according to formal rank, differences concerning opportunities for family life, different salary levels and social status, and so on. Another split was between the mining company employees and the staff at the District Governor's Office; rare cross-over relations and reciprocal scepticism. All these differences and antagonisms have vanished. From a social stratification point of view, contemporary Longyearbyen is an egalitarian middle-class community. The boundaries that exist relate to place of work and profession. 'Where do you work?' is said to be the introductory question when being introduced to a new person.

The next question is 'How long have you been on Svalbard?' and then you are positioned socially. The longer the better: if for a short time, you are at the bottom of the hierarchy. Here time is not only time, but a thick cultural layer. The hierarchy is strongly felt in social interaction. One informant also described a small elite, a network constituted by veterans, people who have been to all corners of Svalbard, who have left their mark on the local community and who are a kind of local celebrities. The hub in the network was said to be a trapper, a former academic, who for several decades has lived alone a few hours' ride by boat or skidoo away from Longyearbyen. Paying this gentleman a visit was among the first things for a new District Governor to do, one informant said, in order to avoid running into trouble. The importance of this network is contested, the informant telling about it is himself part of it. But even if the significance of this network is a myth, it is a strong network that says something about the value of time and identity; these people have a profound sense of belonging to Svalbard.

Altogether, the stability of Longyearbyen is its mobility. It is an open community where newcomers are welcomed, but they have to find their place in a hierarchy based on norms relating to time spent on Svalbard, and partly on the biography people have built up locally. This is a phenomenon known from camp sites where people come to live for the whole summer: your position depends on your performance within the site, not your career or status outside (Angell 1999). When asked to characterize Longyearbyen, the informants applied a series of metaphors. It is like a *folkehøgskole* – a kind of boarding school where youngsters live their entire life for a year or two: a place where people are thrown together, eat and sleep together, undertake common activities and go to the same concerts, restaurants, bars and pubs, and where you make friendships that end some months later. Then after a year or so there is a reunion – on Svalbard this corresponds to the revisits by the ex-Svalbardians. Another metaphor used is *Kardemommeby* in children's literature (written by author Torbjørn Egner) – a make-believe town where people live together in peace and harmony. Yet at the same time, Longyearbyen is 'a small place with an urban lifestyle'. The town is in fact a bit carnivalesque, in the Bakhtinian interpretation of the term – a place for performance, pleasure and play. Even though the town is obviously rooted in 'the eternal and fixed' – well regulated and governed – it is also a reality, characterized by 'ever-changing, playful, undefined forms' of contact (Cresswell 2006, 48).

Mixed Zones and Network Entrances in Longyearbyen

How then, do people become integrated in Longyearbyen? Some of the informants explained that they had relatives or friends living in town. In these instances, the way into the community is easier. However, most people have no such relations. A second way to be brought into the community is by being introduced to somebody – most often by a colleague with some Svalbard experience. These are important middlemen. Informants say that they often speak to newcomers to find out what interests they have and tip them off about other people and arenas for these activities. One of them said:

> I recently found out that a new guy used to play indoor hockey, and knew that there was a group playing this sport regularly; now he is part of that group. Another example was a former speed-skater (on ice) who was here on a three-year contract. He gathered people to make a track, and many people bought skates and engaged in the sport. However, after he left, the activity stopped. Thus, some activities rely on enthusiasts. I myself engage in golf. Up here this is a kind of extreme sport.

Another lady said that at first she was introduced to people through 'welfare arrangements' at the District Governor's office, where her husband had got a job, 'but this was only the first week, then I started with five activities, and after two weeks I knew more than 100 people'. Thus, a third bridge to the social network is to take part in organized activities and become a member of one of the many sociations. There are more than fifty local NGOs, more or less active and more or less organized. Many of these associations seem to be rather broad social arenas: they organize tours (by boat or skidoo), their members go together to pubs and bars and they orchestrate social happenings and Christmas parties. It was claimed that these sociations are far more social than would normally be the case on the mainland.

A fourth arena of integration consists of the kindergartens and the school, which are of course particularly important for people with children. Activities here involve parents and they are more or less forced into contact and collaboration. The school is also a cultural site and where the sports hall is located, and thus where people gather. In addition, some take part in running these facilities as arenas for leisure activities or in the organization of cultural events and activities, and some are involved in the management of the school. Consequently, people merge.

Mixed zones are places where people with different roles – in sports terms, athletes and journalists – are supposed to interact. This is where strangers have a fair chance to get acquainted, and here newcomers meet the veterans. The most common mixed zones are restaurants and bars, but public places, streets and institutions such as museums, art exhibitions, galleries and shops may also serve this function. These are places where you are not supposed to have to qualify for access. However, in Longyearbyen only the bars and pubs seem to be places for public encounters, and more for young people than for others. Nevertheless, people in Longyearbyen are outgoing, far more than is the case in mainland Norway, and are more open to making new acquaintances, according to an informant comparing Longyearbyen with Tromsø, the nearest sizeable town on the mainland. For a few, these bars and restaurants are major social arenas. As a female Svalbard veteran said, 'I can easily go out alone; there is always someone to join.' It is a contact zone (Pratt 1991), but not an important place for knitting social networks, it was maintained, more a place for transient encounters. However, it was maintained that 'this arena was more important before, when people were accommodated in very small rooms – you had to go out to breathe'. Then there were also places where different cultures met, for instance those of tourists and miners. The diminished importance of such places probably also relates to a more global trend hitting Longyearbyen, the tendency towards a vanishing of public sociability (Wellman 1999, 28). Another way of interpreting Longyearbyen in this instance is to say that sociability has moved from public places to semi-public sociations.

Is There a Local Identity?

The accounts above tell of social networks relating to jobs, the school and numerous sociations, and about relations with people living in other places. Most of the informants obviously have strong affiliations to the place where they used to live, or were they were born, and they frequently return – one lady said about ten times a year – and many intend to go back some time in the future. Since Longyearbyen is a place that most people move to for a period of time, there is no reason to expect that the ties to the place should be particularly strong. Many express a sense of identification with Svalbard, but have problems expressing what the local identity is, or whether they really identify with the town. However, as a Svalbardian you acquire an image that is attached to you, relating to the remote location, the climate and the exoticism of the archipelago. In social settings outside Svalbard people have had the experience of becoming a focal point socially – there is an obvious aura relating to the role of a Svalbardian. However, many regard this as strenuous – communication often takes the form of an interview, with people asking questions about the islands and life up there, blocking normal social chat. This is not a chosen modus, but something ascribed to an individual with a connection to Svalbard. Thus, the identity relating to Longyearbyen is at best related to an individual's profession. For most people Svalbard, but also Longyearbyen, is an aspect of the job they perform. Few people have jobs relating to a product for the external market.

There are strong social rules concerning who one should not identify with in Longyearbyen: people with a short Svalbard horizon, and tourists. For the locals, even for those on a short-term stay, it is important to keep a distance from tourists who have to 'buy' their experiences. Transactional relations are perceived as less valorized bonds. It is a local 'sport' to make jokes about tourists, and it is important not to behave like them. Often the tourists are rather visible, with all their clothes, cameras, strolling around in groups and often carrying rifles. Locals 'have flour in their shopping trolleys and never carry a gun in town', it was maintained. One way to be accepted seems to be to emphasize bonds to people within the community and mark the differences compared to the outside world. Thus, there is obviously a tendency to create communality by expressing distinction from others (cf. Stedman 2002) and from the sphere where commercial relations reign.

Conclusion

This chapter has been looking at a community with an extreme turnover. However, the mobility is the normality – in many different ways – and people seem to have learned how to cope with this situation. One of the interesting findings is that since people do not normally have kinship relations on the islands, friends seem to fill some of the functions of kin, with regard to both aid exchange and intimacy. Another is that friends and acquaintances are often also workmates. In this respect Longyearbyen is an extreme, although socializing with workmates is said to be one aspect of the post-modern society (Rojek 1995, cf. Savage et al. 2005). On the other hand, a spatial split between industrial and residential areas is another trait of contemporary society, but this is not the case in Longyearbyen. Here people live close, but work nevertheless obviously generates the most important networks and friendships seem to be situational: many of them end when people leave. Somehow there seems not to be room for what Giddens (1991) has called pure relationships: friendships developed through being brought up together, having got to know each other through third parties, and so on.

Another tendency is that for many people, their best friends live far away. This is an additional aspect of networks in Longyearbyen: being dispersed, extending out of town and maintained by the use of communication technology and travel. It is consistent with findings in other places. "Best friendship' can be seen as an abstract social relationship that takes people out of the lived routines of the daily life, and which links them to spatially dispersed others', say Savage et al. (2005, 151). Closely related to this, a final major finding is that Longyearbyen is a community where people are involved in a variety of networks, both of local and non-local character. This is in accordance with the general pattern that Wellman (2001) has observed. Thus, it is not astonishing that it is difficult to say what the Longyearbyen identity is – if it exists. The town is most probably not a communal identification.

The overall conclusion is that one of the most mobile communities in Europe is a good place to live for those who can sustain the first period, and for people interested in nature and outdoor recreation; if not, there may be problems. It is a limitation of this study that only those who have overcome the first period have been interviewed – in itself a sign of adaptation. Those who have not adapted to the local customs, norms and values are not living there. Every society of inclusion may have an excluding side. So also on Svalbard. Longyearbyen is an artificial community, upheld by subsidies and Norwegian sovereignty politics. Thus, it is a privilege of this place not to bother about those who do not fit within the local norms and normality.

Acknowledgement

I am thankful for valuable discussions with Tove Eliassen, and for her preparation of and collaboration concerning the focus group interviews.

References

Angel, E. (1999), 'Campingplassen og modernitetens nomader', in Jacobsen and Viken (eds).
Arlov, T. (1996), *Svalbards historie* (Oslo: Aschehoug).
Baldachino, G. (ed.) (2006), *Extreme Tourism. Lessons from the World's Cold Water Islands* (New York: Elsevier Science).
Bauman, Z. (2001), *Liquid Modernities* (London: Sage).
Bourdieu, P. (1977), *Outline of a Theory of Practice* (trans. R. Nice) (Cambridge: Cambridge University Press).
—— (1986), 'The Forms of Capital', in J. Richardson (ed.).
Castells, M. (ed.) (2004), *The Network Society. A Cross-Cultural Perspective* (Cheltenham: Edward Elgar).
Clifford, J. (1997), *Travelling Culture* (London: Routledge).
Coleman, J.S. (1988), 'Social Capital in the Creation of Human Capital', *American Journal of Sociology* 94, 95–120.
Cresswell, T. (2006), *On the Move: Mobility in the Modern Western World* (New York: Routledge).

Ferrand, A., Mounier, L. and Degenne, A. (1999), 'The Diversity of Personal Networks in France: Social Stratification and Relationtion Structures', in Wellman (ed.).

Giddens, A. (1991), *Intimacy and Self-identity* (Cambridge: Polity Press).

Glover, T.D. and Hemingway, J.L. (2005), 'Locating Leisure in the Social Capital Literature', *Journal of Leisure Research* 37, 387–401.

Hemingway, J.L. (2006), 'Leisure, Social Capital and Civic Competence', *Leisure/Loisir* 30, 341–55.

Hetherington, K. (1997), *The Badlands of Modernity: Heterotopia and Social Ordering* (London: Routledge).

Jacobsen, J.K. and Viken, A. (eds) (1999), *Turisme. Steder i en bevegelig verden* (Oslo: Universitetsforlaget).

Kearney, M. (1995), 'The Local and the Global: An Anthropology of Globalisations and Transnationalism', *Annual Review of Anthropology* 24, 547–65.

Larsen, J., Urry, J. and Axhausen. K.W. (2007), 'Networks and Travel: The Social Life of Tourism', *Annals of Tourism Research* 34, 244–62.

Longyearbyen Community Council (2006), *Samfunns- og næringsanalysen Svalbard 1991–2005* (Longyearbyen: Longyearbyen Lokalstyre). http://www.lokalstyre.no/Modules/theme.aspx?ObjectType=Article&ElementI=765&Category.ID=710, accessed 30 May 2007.

Macnaghten, P. and Jacobs, M. (1997), 'Public Identification with Sustainable Development. Investing Cultural Barriers to Participation', *Global Environmental Change* 7:1, 5–24.

Morgan, D.L. (1996), 'Focus Groups', *Annual Review of Sociology* 22, 129–52.

Pawar, M. (2006), 'Social Capital', *The Social Science Journal* 43, 211–26.

Pratt, M.L. (1991), 'Arts at the Contact Zone', *Profession* 91, 33–40.

Putnam, R.D. (2000), *Bowling Alone. The Collapse and Revival of American Community* (New York: Simon & Schuster).

Rapport, N. and Dawson, A. (1998), 'Home and Movement: A Polemic', in Rapport and Dawson (eds).

—— (eds) (1998), *Migrants of Identity. Perceptions of Home in a World of Movement* (Oxford: Berg).

Richardson, J.G. (ed.) (1986), *Handbook of Theory and Research for the Sociology of Education* (New York: Greenwood Press).

Rojek, C. (1995), *Decentering Leisure Theory* (London: Sage).

Savage, M., Bagnall, G. and Lunghurst, B. (2005), *Globalization and Belonging* (London: Sage).

Statistics Norway (2007), http://www.ssb.no/emner/02/02/20/flytting/, accessed 15 March 2007.

Stedman, R.C. (2002), 'Toward a Social Psychology of Place – Predicting Behavior from Place-based Cognitions, Attitude, and Identity', *Environmental Behaviour* 34:5, 561–81.

Urry, J. (2000), *Sociology beyond Societies. Mobilities for Twenty-first Century* (London: Routledge).

Viken, A. (2006), 'Svalbard', in Baldachino (ed.).

Wellman, B. (1999), 'The Network Community: An Introduction', in Wellman (ed.).
—— (ed.) (1999), *Networks in the Gobal Village. Life in Contemporary Communities*
    (Oxford: Westview Press).
—— (2001), 'Physical Place and Cyberplace: The Rise of Personalized Networking',
    International Journal of Urban and Regional Research 23, 228–52.
Wellman, B. and Berkowitz, S.D. (eds) (1988), *Social Structures. A Network
    Approach* (New York: Cambridge University Press).

PART 3
Mobilizing Place

Chapter 13

Enacting Places through the Connections of Tourism

Gunnar Thór Jóhannesson and Jørgen Ole Bærenholdt

Introduction

Tourism has for a long time been to the fore as a prominent feature in the discourse on socio-economic development. This applies not least to places in the periphery (for example, Crick 1989; Hall, Roberts and Mitchell 2003; Shaw and Williams 2004). The idea of tourism development implies that the place where it is carried out changes. The development process is (for better or worse) intended to affect the general enactment of the place in question. This chapter addresses the process of place enactment in relation to a particular tourism development project carried out in the village of Thingeyri, Iceland. We identify spatial ordering of relations and political practices as two fundamental features of place enactment.

Our theoretical point of departure is a relational approach, much inspired by actor-network theory (ANT). This perceives places as being constituted by intersecting connections of past, present and future-oriented mobilities and moorings. From such a perspective, mobility and place cannot be thought apart from one another. Connections are understood as being, in the most basic terms, manifestations of space-producing practices. As such, connections are integral to the enactment and existence of place. Without enacted connections, the place would simply not 'happen'. This view disturbs the idea of a distinction between the local and the global, and breaks away from thinking of the local place as having territorial boundaries that define its inside and outside. We do not pretend that there are no boundaries, but we argue that these have to be seen as the effect of spatially diverse connections driven and shaped by political practices. While some are obviously local or territorial, places in the periphery are also enacted through far-reaching and transnational connections.

This chapter explores parts of this process by describing the ways in which people in the village of Thingeyri in Iceland have worked on a transnational tourism development project entitled 'Destination Viking Sagalands – Sagas and Storytelling'. The project is primarily financed by the European Union's INTERREG IIIB Northern Periphery Programme (NPP), and as such is an example of the transnational connections of tourism mobilities. This is an example of tourism development projects that are intended to (re)enact places into attractive environments, not only for tourists but also for the locals. To understand place enacting, we contend that it is crucial to attend to the very concrete spatialities and politics involved in the work through which particular projects are being accomplished.

The chapter starts with a conceptual discussion, framing the rest of the chapter. It then recounts the translation process of the Destination Viking Sagalands project to Thingeyri. The idea of translation refers both to movement and to transformation, and it is demonstrated how the Destination Viking Sagalands (DVS) project undergoes important changes on its way from the original idea put forward in the project application to the Northern Periphery Programme (NPP) to this being put into practice in the village of Thingeyri. We then illustrate how this process casts light on place enactment, with a special focus on the spatialities and politics involved. At the end we conclude that the apparent fluidity of the spatial ordering of the DVS project is of considerable significance in the accomplishment of the project in Thingeyri.

How are Places Enacted?

A relational approach to place enactment takes its cue from a general practice-based ontology. The key concepts of mobility and place are understood as effects of spatial and temporal practices. With reference to actor-network theory (Latour 1993; 2005; Law and Hassard 1999), we contend that mobility and place are enacted through people's practices in association with an array of non-human elements. Hence, mobilities and places are beyond 'pure' human performance: they depend on the materiality of nature and various technological systems, such as cable connections and air transport systems.[1] These act as stabilizing moorings, which afford certain practices while excluding others (Bærenholdt et al. 2004; Hannam et al. 2006) (see also Chapter 8 by Larsen and Urry).

Place may be understood with the help of Doreen Massey's (2005) notion of 'throwntogetherness', which vividly depicts how places are spatio-temporal events (Massey 2005, 130) (see also Chapter 2 by Simonsen). The idea of 'throwntogetherness' highlights how place and mobility are intrinsically entwined. 'Throwntogetherness' necessarily happens through some sort of mobility, and at the same time mobility can be a form of 'throwntogetherness'. Places are constantly in the making, for instance when people travel together, passing along the same routes, relying on the same moorings, sharing certain forms of obligation or responsibilities that constantly produce a series of encounters through time and space (Bærenholdt 2007). Mobilities, such as those manifested in transnational tourism connections, thereby add to the production of the socio-spatial landscape that constitutes a place.

Tourism connections, like other mobilities, do not enact places in a vacuum. They draw on and add to various material, social and cultural relations that are more or less settled. In order for tourism development projects to succeed, they have to relate to this sedimented mixture of spatio-temporal events. Place enactment and re-enactment is a manifestation of a continuous construction, ordering and testing such relational configurations – a process aptly described by Massey as 'negotiating a here-and-now' (2005, 140). This ongoing process of negotiation includes efforts to construct or make connections to new or alternative mobilities.

---

1   The message is that agency and practice are always hybrid. One cannot think about human practice as an isolated achievement: it is accomplished through heterogeneous associations.

Connections can be made to networks more enduring and regularized than just a single event or encounter. This is the case in Thingeyri, where the Destination Viking Sagaland project (hereafter DVS) relates elements of tourism, regional politics, businesses and civic engagement across national borders. The DVS is an example of a strategic pursuing of certain objectives, which in general can be framed as regional development in the periphery. It is an assemblage that drives the development of tourism in the village, and as such it is a clear case of place enactment. This immediately calls attention to the political side of the process, invoking questions in relation to establishing connections, choosing path(s) to follow or shaping a development strategy to pursue.

It is evident that for many transnational development initiatives, such as the European Union projects supported by the Northern Periphery Programme, the creation of transnational networks, perforating nation-states, is an important objective:

> The overall aim of the programme is therefore to encourage joint projects that address the priorities for co-operation shared by the participating countries. Through the *exchange of knowledge and experience* it is anticipated that some of the barriers to a more *balanced development* of the area can be overcome. (Northern Periphery 2004, emphasis added)

The networks that configure the 'exchange of knowledge and experience' are indeed expected to contribute to the making of a more 'balanced development' in the regions in question. Yet between the lines one may also infer that the networks shaped through this transnational programme are in fact more than instruments for the implementation of projects: they are more or less explicitly ends in themselves. Thus, the programme aims partly at a particular form of spatial ordering of (project) relations.

In general, the quotation above is an expression of what Ash Amin (2004) calls the politics of connectivity. It is a concept that is intended to grasp aspects of relational politics, able to grapple with a political space that does not necessarily match the grid, like the surface of Euclidian/regional space. Amin takes as his point of departure the 'spatial stretching and territorial perforation associated with globalization' by way of international transport and communication, but also 'faith communities, dream worlds and cultural domains that cut across lines of longitude and latitude' (Amin 2004, 33). Even so-called peripheries are becoming places imprinted by the mobility of both visitors and inhabitants to some extent, where one might pose the question: who is who? Politics of connectivity refers to an environment where public life is made up of '…varied geographies of relational connectivity and transitivity…' where it is '…open to both local and distant actors to sign up to a given programme…' (Amin 2004, 40, 41).

Amin combines politics of connectivity with politics of propinquity '…shaped by the issues thrown up by living with diversity and sharing a common territorial space' (Amin 2004, 39). It cannot be stressed enough that these concepts should not be interpreted as a dualism, reproducing a well-worn distinction between the bounded territory/community and the mobile/global/de-territorialized society. There is an underlying element of tension, between (co-)presence and absence, or proximity and distance. The politics of place enactment revolves around the ordering of the relation between proximity and distance into various spatial forms of practice (see below). For instance, politics of propinquity have to do with issues thrown up by co-presence or 'nearness' in a given territory, that is, a regional space. Other political practices, taking place far away, may matter to the place, but they do so not because of their present regional weight (their propinquity) but rather because of their more network-type or fluid-like connectivity. Their significance is felt only when they become connected to a particular place, thus affording them an absent presence. Absent in the propinquity of regional space, but present in the connectivity of network/fluid space.

It follows that the politics of propinquity and the politics of connectivity are, in our view, intrinsically entwined, providing a first and foremost heuristic device to explore what we call the ontological politics of place enactment. Ontological politics suggests that reality is

> ...*done* and *enacted* rather than observed...reality is manipulated by means of various tools in the course of a diversity of practices. (Mol 1999, 77, emphasis in original)

Moreover, the idea of ontological politics implies a '...link between the real, the conditions of possibility we live with, and the political' (Mol 1999, 86). Ontological politics are thereby concerned with enacting different versions of reality, relating them, manipulating them, holding them apart, turning a potential into a reality or making some versions of reality more real than others (Law 2004). In brief, we state that the ontological politics of place enactment revolve around (the contested issue of) making things present.

The Translation of the 'Destination Viking Sagalands' Project to Thingeyri

We shall now discuss specific moments in the process of moving the DVS project to the village of Thingeyri. This is a process of translation that highlights the somewhat messy spatialities and politics involved in the enactment of place through the transnational connections of tourism.[2]

The DVS project began with meetings between Rögnvaldur, a tourist consultant based in Reykjavík and associated with Iceland's Institute for Regional Development, and the Nordic Atlantic Co-operation (NORA), an agency under the Nordic Council of Ministers, its central office located in Tórshavn in the Faroes. NORA established connections with the Northern Periphery Programme (NPP), based in Copenhagen, and with a number of other international entrepreneurs who had been involved in earlier EU INTERREG projects. Based on these and many other contacts across the

---

2   The case is more thoroughly explained in Gunnar Thór Jóhannesson's monograph (2007). Ethnographies were written up as part of a broader contribution to the development of an actor-network approach to tourism economies, building especially on concepts and metaphors from actor-network theory (ANT), such as translation, regions, networks, fluids and fires (Jóhannesson 2005a, 2007), adding these ideas to a discussion of the politics of connectivity. The following account is based on parts of Chapter 6, 'Networking Tourism Economies'.

Atlantic, Rögnvaldur applied for funding from the NPP, which is firmly embedded in the EU's transnational regional policies. The project synopsis refers strongly to sagas and storytelling as unique features of the northern periphery and argues that the revitalization of these traditions could form steps to attain objectives of the NPP, such as community development and social inclusion.

It took many meetings to construct and extend the project network to various partners. In the end it stretched even beyond the immediate boundaries of the European periphery, as demarcated by the EU. Partners came from Newfoundland, Greenland, Iceland, the Faroes, Scotland, the Isle of Man, Norway and Sweden, and Rögnvaldur acted as project manager. We shall now focus on how the project was translated from the state of project application to the specific actions taken in the village of Thingeyri, in the Westfjord region of Iceland.

In general, this is a story of making connections. First, here is a glimpse of some of the traces that were gathered into the project. Some have to do with Dorothee Lubecki (called Dóra). She is the tourism consultant for the Westfjord region, based in the administrative centre of the region, the town of Ísafjördur. She came originally from Germany and has been working on tourism development in the Westfjord region since 1996. Soon after she started her job as tourist consultant she experienced at first hand, while driving with a local farmer, how vivid the Saga of Gísli Súrsson was to some of the people in the area. At the time she had the idea that the Gísla Saga could be used to develop tourism around Thingeyri, which is where most of the saga takes place. Though absent and dead for almost a thousand years, Gísli Súrsson seems somehow to be 'alive' and not that distant at all. He is present in the cultural landscape of the area, as well as in the imaginations of both the locals and many other Icelanders, since his saga has been compulsory reading in schools for decades. Then along came a German student, who had travelled to the Westfjords and written a master's dissertation on the possibilities of developing Saga-based tourism in the area, focusing especially on the Gísla Saga. Dóra brought some of the ideas from her dissertation to the fore at a 'community' meeting in Thingeyri as early as 1999, but this did not result in any concrete action.

There are other actors relating to the story, too. One is Thórir, an electrician living in Thingeyri; he got his diploma as a local guide back in 1994 and soon began to take tourists to the sites of the Saga. Another important actor is Thórhallur, who moved to Thingeyri in 2000. He was connected to the place through family relations – his grandparents had lived there – but he was now moving there for professional reasons associated with the fishing industry. Last but not least, Gísli Súrsson is still alive through the Saga and the many associated topographical names.

Now, let us go back to the beginning of the DVS project. Rögnvaldur contacts Dóra in 2002. At the time, the only connection to Thingeyri is the absent presence of Gísli Súrsson and the German master's dissertation lying on a shelf in Dóra's office in Ísafjördur. Dóra makes the connection between the tourism development project that Rögnvaldur is trying to assemble (DVS) and the potential of the Gísla Saga. She adds her input to the overall project and then begins to introduce the idea to people in Thingeyri. She starts a course in weaving, and little later calls for a meeting in Thingeyri to introduce the project and see whether there are people who would like to sew Viking costumes for themselves. She is well aware of the handicraft traditions among the women in the village, skills that were also mobilized in response to a crisis in the fisheries years before (see Skaptadóttir 1998). Around 30 people begin to sew Viking clothes and Dóra organizes a brainstorming group in connection with the course. It is very important that Dóra is not too bound to the overall project document for the DVS, to which the Northern Periphery Programme has now granted money. Instead of maintaining the project's shape (on paper), she invites people to take part in shaping it in practice. The project thereby becomes mutable.

The brainstorming group present their ideas in a report published in 2003 and suggest the formation of a local non-profit organization, the Westviking Association, which would draw on the DVS project but also work on general tourism development in the area. Later that same year, the Westviking Association is formally established at a meeting in the community house. Among the board members are Dóra, Thórhallur (chair) and Thórir (cashier). An enactor of tourist places in the Icelandic periphery has been constructed. Among the effects of the actions of the Westviking Association are the construction of the Viking Ring, outdoor festival facilities located on the outskirts of Thingeyri, made of turf, timber and stones. Furthermore, much work has been put into the making of signposts to be positioned by the side of the road running through the central scenes of the Gísla Saga (Jóhannesson 2005b; 2007, Chapters 7 and 8).

Much of the energy put into these activities derives from the engagement first concentrated in the very material practice of sewing and weaving, which literally connected local people together and created a sense of fellowship among the participants. It is important, though, not to forget older traditions and the vivid heritage of Gísli Súrsson, brought together by a few key actors, who metaphorically initiated the weaving of the project into being by pushing for action, applying for funding in addition to the funding from the Northern Periphery Programme, and carving out an agenda for the work of the Westviking Association.

The story about the movement of the DVS project to Thingeyri implies both the translation from the Northern Periphery Programme to the 'region' of Thingeyri and the translation from the state of immutable mobile, in the form of a project paper, to lived practices. Both aspects of the translation happened through 'networking'. The next two sections deal with the spatialities and politics of this process of networking, which manifest place enactment in Thingeyri through the connections of tourism.

Spatialities of Enactment

The translation process of the DVS project to Thingeyri evokes three forms of spatial ordering or topology which, following Mol and Law (1994), may be described using the metaphors of the region, the network and fluidity. The first two are commonly recognized. The region refers to Euclidian space that depicts the world as a grid-like surface on which it is possible to draw durable lines of metric distances. The 'network' refers to such lines that connect nodes, which may be situated in geometric space. When these forms intermix, they allow for the movement of immutable mobiles between places. One example is the project application of the DVS. It was moved to Thingeyri through a stable network connection (email) that stretches across Euclidian space. The central feature of the network, however, is that as long as all its working components hold their posts it affords action at-a-distance as it remains the same in different places (see Law 1986). In the case above, this function of the network seems to break down, since it soon becomes evident that the DVS project is no longer the same in Thingeyri as it was in the project application. The networks of reciprocity at-a-distance do not hold in a consistent fashion, and in turn we need to look for alternative spatial imaginaries to explain how something still happens in Thingeyri. This leads us to the metaphor of fluidity.

By implementing the project in Thingeyri, Dóra renders it open to change, encouraging alterations. Here the translation of the project enacts another kind of space, which the metaphor of fluidity best grasps. Bearing in mind that we are dealing with practices and the diverse space-times that they produce and happen through, the topological form of fluidity catches the ways in which relational continuity is sustained through change (Mol and Law 1994; Law and Mol 2001). The DVS is carried out and made firm in Thingeyri by way of making the project mutable. The shape-shifting opens up a space for the engagement of a number of local people, but this also results in some confusion. One might ask if this is still the same project? The answer is that it depends. Dóra tried to manage this play between the immutable EU network and the mutable, fluid connections through which it was enacted in Thingeyri and beyond. More often than not, she succeeded. The most important point is that it is most probable that nothing would have happened at all if she had not made it available for change through the practices of the local people.

The metaphor of fluidity for the mutability of the DVS project conveys an idea of the crucial spatial and temporal messiness involved in the realization of the project in Thingeyri. It refers to a spatial form of practice that has been largely neglected in studies of regional development. We suggest that the spatial form of relational ordering that is embodied by fluidity is, rather than an abnormal instance, ordinary in project work, especially in projects that actually succeed in some form of implementation. Often the non-messy is not at all enacted, as it is simply too fragile to survive the journey of translation. In other words, in a changing world mutability equals strength.

The three types of spatial forms of relational enactment that we have discussed should only be taken as ideal types of relational configuration. They describe different forms of 'spacing' and 'timing', and cannot be thought of as being independent of practices. Thus, these forms do not lead a life of their own, but are expressions of the spatialities and temporalities of practices themselves. The translation of the DVS project, understood as an enactment of transnational mobilities, happens through relational practices that draw on and add to interfering forms of spatialities. The practices studied also have their politics, to which we shall now turn.

Politics of Enactment

Above, we framed politics of enactment as being ontological politics, where the ordering of the relation between proximity/presence and distance/absence was central. Significantly, the idea of ontological politics provides a link between forms of spatial ordering of relations to the substance of practices; practices that relate to the initiating, assembling and ordering of relations. The political issues at stake, in the most basic terms, have to do with what kind of reality should be pursued, what should be made present, what should be kept absent and how to succeed in doing this. In the present example, the most critical issue for the successful establishment and translation of the DVS project was therefore to gain (political) support, both from the NPP and from the inhabitants of Thingeyri. From a relational perspective to politics, the power of the DVS as a development project rests on the confidence that different actors have in the project.[3] These incorporate some of the key actors (Dóra, Thórir, Thórhallur) and members of the wider public (participants in the sewing course). In order to explore this we might usefully turn again to Amin's concept of the politics of connectivity and politics of propinquity.

It is evident that the creation of the DVS project was in many ways an expression of the politics of connectivity, where actors signed up to common programmes and formed a kind of mobile bond at a distance. Though these people may have met by coincidence, this was not just something that happened out of a clear blue sky. Enactors chose their fellow actors and enroled them into stable networks for strategic reasons, primarily to 'catch' EU and other funding in order to make their project economies work. However, there were already important connections in place from the past, which until being gathered together only hinted at the potential involved in projects like the DVS. In the case of Thingeyri, these included Dóra in Ísafjördur, Thórhallur and Thórir in Thingeyri, the German master's dissertation leading a shelf-life in Dóra's office, her informal talks with locals in the area about the potential of Gísla Saga for tourism, and of course Gísli's absence-presence. These are the central threads, which when gathered have driven the enactment of the DVS project in Thingeyri and thus the enactment of Thingeyri as a place of tourism in the making. Yet in order to make it 'happen' there has been more at stake than the complex politics of connectivity, spanning the project enactors' strategic way of 'throwing' particular people together and the many traces coming together in Thingeyri.

This 'something more' aspect has to do with civic engagement and commitment among the people belonging to Thingeyri, a village that has previously been hit by crises and where people have tried to cope in several ways. The emergence of civic engagement, expressed in the sewing course and the project development work that ensued, is a moment of politics of propinquity, stemming from the 'throwntogetherness' of people in this place (see above). In spite of many people moving in and out, and the possibility of leaving the place via the local airport or the roads, either northwards towards Ísafjördur or southwards towards Reykjavík, living, having a house and maybe even working in the village have important implications. They imply a certain destiny of having been thrown together in a particular place (Massey 2005), and thus of sharing present circumstances of everyday life.

As mentioned before, we hold that politics of propinquity and politics of connectivity are usually entangled together in such a way that it is difficult to set them apart. As Amin notes, everyday (local) life is constituted through distantiated attachments and influences as well as through sharing a place (2004). The politics of place enactment thus have to do with all the connections that somehow affect a

---

3   It is therefore associational (Latour 1986), rather than being a property of individual actors or bodies of organizations.

given place. The account above provides examples of practices that relate to both connectivity and propinquity, but there are also other aspects of the process.

One of these, of course, is the more strategic discourse and initiatives taken by municipal and development agency actors in enacting local economic development. These express a full-blown version of the territorial politics of place, which perceives a place or region as a territorially-demarcated whole that can (and should) be subject to general socio-economic management (Amin 2004). This is a discourse that may or may not become connected to specific places in the periphery; in the case of Thingeyri it was. The clearest example is the role the Westfjord Development Agency played in the whole process as Dóra's employer. The agency works on the basis of this kind of discourse of regional development, heavily promoted by national authorities (Byggðastofnun 1999; Alþingi 2002). Another example is the way in which the Westviking Association sought to connect to pools of capital intended for regional development (cf. detailed discussion in Jóhannesson 2007, Chapter 8). But these institutions of regional development would not work without the kind of engagements we have described above.

Another issue is the mobile contingencies, which have not yet been studied to any great extent. To enact a place as a tourist location fully really requires the performance of tourists (Bærenholdt et al. 2004). The Westfjords, far away from Iceland's ring road, are not the most visited tourist destination in Iceland, especially not on the part of international tourists. In 2004, the region accommodated only around 1 per cent of registered overnight stays by foreign tourists and 3 per cent of registered overnight stays by Icelanders (Guðmundsson 2004). Air connections can be uncertain, especially during the winter, and the roads are long and notoriously troublesome. Nevertheless, tourism in the region is very much car-based, with more than 90 per cent of Icelandic tourists arriving in their own cars and more than half of foreign visitors coming by (often rented) car (Atvinnuþróunarfélag Vestfjarða and Ferðamálasamtök Vestfjarða 2004).

It is therefore evident that some of the automobile tourists who drive through the Westfjords will pass Thingeyri. This banal fact is of crucial importance. While the making of tourist places may have a lot to do with connections and civic engagement on the part of the locals, this does not help if tourists are not passing through. Some visits may be rather networked, such as those by family, friends and re-visitors heading to a specific 'destination', but other connections of tourism are more fluid or even coincidental, depending on circumstances inside and outside the car, not least the weather. Signposts re-enacting Gísla Saga, for instance, are attempts to catch passing tourists and convince them to take part in enacting a tourist place by sharing 'local space', even if only in passing.

Here, some points hint at political practices of connectivity, whilst others relate to elements of propinquity – practices that depend on nearness, that is, sharing a space. Place enactment happens through complex political practices taking diverse spatial forms. As we shall see in the final section, tourism development in the northern periphery partly depends on allocating space and time in order for this spatial and political messiness to evolve (see also Jóhannesson 2007, Chapter 9). Mutable connections are crucial, whereas strategies that are distilled into the purified forms of regions and networks hardly suffice in coping with the unpredictable events of the present era.

Mobility and Place

This chapter has shown that places in the periphery, such as Thingeyri, are enacted in multifarious ways. The place of Thingeyri is best understood as an arrangement of relations that shape its socio-spatial configuration. As such, it is porous and – as the inhabitants have experienced through socio-economic crisis – it may also be a fragile constellation. Yet, situated in (topological) geographies of distanciated relations, they are able to bypass conventional network politics and ways of thinking in regional development discourse in surprising ways.

Firstly, the enactment and transformation of the periphery implies diverse spatial and temporal orderings. The traditional network approach, based on a reciprocal relation between a few actors (whether in business or not), can hardly explain the types of dynamic change that we have observed. We have pointed towards a fluid form of spatial ordering of transnational mobilities, no less important to socio-economic innovation than networks. The mobilities involved in the DVS project are not only about 'transporting' an idea, an object or a person from 'A' to 'B'. If it were not for the fluidity of some of the connections, that is, the mutability of ideas, objects, aims and opinions, not much would have happened. It is imperative that local and regional development is flexible enough to cope with such fluidity, which is generally driven by the diversity of actors and connections involved in such agendas.

Secondly, the discussion has underlined how place enactment is a political process. The connections and propinquities involved in the practices of enactment must be identified as varied spatio-temporal processes, which should not promote a local-global juxtaposition. Their exact expression has to be studied in each case. Practices are always somehow localized. At the same time, actors can build their relations at-a-distance and stabilize this way of working through the sense of trust, fun and engagement derived by meeting face-to-face from time to time. In this way, mobilities and places intersect in complex ways, which may even undermine the distinction between these two concepts.

It is important that the relational approach we have suggested should not distract attention from questions of inequality and the processes of inclusion and exclusion. Instead of thinking about these processes in terms of 'regional' geography, we contend that it is much more effective to trace the enactment of connections that create lines of distinction among people in practice. That is, we should attend to the practical implications of the continuous negotiating of here-and-now; the continuous and always political process of place enactment does not respect the regional boundary of place, but neither does it abstain from taking its territorial significance into account. Place does not equal cultural homogeneity, nor does mobility equal heterogeneity. People may well bridge diverse identifications by being thrown together in place, just as people may bond around common identifications through shared mobile practices (Bærenholdt 2007). As we have seen, this means that the ontological politics of place enactment are messy, although they can sometimes be meaningfully gathered under the umbrella of 'connectivity' and 'propinquity'.

Yet it is exactly this messiness that allows for creativity, since it does not necessarily predefine which practices and which motivations matter in the process of making things present. We have argued that if practices of change are not somehow connected to the everyday life of people they will often not accomplish any significant social change. Civic motivation may be absolutely decisive to enactment, and this involves certain forms of fellowship among people, often stabilized through material elements such as the Viking clothes that the people in Thingeyri sewed together.

Flexibility and openness, together with engagement and commitment, are crucial energies for change in Northern European peripheries. This underlies the vitality, but also the fragility of the multifarious ways of life practised, and thus the enactment of place.

References

Aarsæther, N. and Bærenholdt, J.O. (eds) (1998), *Coping Strategies in the North, Local Practices in Context of Global Restructuring*, INS 1998: 303 (Copenhagen: MOST and Nordic Council of Ministers).

Alþingi (2002), *Tillaga til þingsályktunar um stefnu í byggðamálum fyrir árin 2002–2005* (Reykjavík: Alþingi).

Amin, A. (2004), 'Regions Unbound: Towards a New Politics of Place', *Geografiska Annaler* 86B: 1, 33–44.

Atvinnuþróunarfélag, V. and Ferðamálasamtök, V. (2004), …Náttúrufegurð, kyrrð, gott mannlíf og blómstrandi menningarlíf sem stendur á *gömlum merg* …*Niðurstöður könnunar á meðal ferðamanna á Vestfjörðum sumarið 2003* (Ísafjörður: Ísafjarðarbær).

Bærenholdt, J.O. (2007), *Coping with Distances: Producing Nordic Atlantic Societies* (Oxford: Berghahn Books).

Bærenholdt, J.O., Haldrup, M. and Larsen, J. (forthcoming), 'Performing Cultural Attractions', Darmer, P. and Sundbo, J. (eds), *Production of Experiences*.

Bærenholdt, J.O., Haldrup, M., Larsen, J. and Urry, J. (2004), *Performing Tourist Places* (Aldershot: Ashgate).

Byggðastofnun (1999), *Byggðir á Íslandi: Aðgerðir í byggðamálum* (Sauðárkrókur: Byggðastofnun).

Crick, M. (1989), 'Representations of International Tourism in the Social Sciences: Sun, Sex, Sights, Savings, and Servility', *Annual Review of Anthropology* 18, 307–44.

Guðmundsson, R. (2004), Ferðamenn á norðanverðum Vestfjörðum. Rannsóknir og ráðgjöf ferðaþjónustunnar (Reykjavík: report).

Hall, D., Roberts, L. and Michell, M. (eds) (2003), *New Directions in Rural Tourism* (Aldershot: Ashgate).

Hannam, K., Sheller, M. and Urry, J. (2006), 'Editorial: Mobilities, Immobilities and Moorings', *Mobilities* 1:1, 1–22.

Jóhannesson, G.T. (2005a), 'Tourism Translations: Actor-Network Theory and Tourism Research', *Tourist Studies* 5:2, 133–150.

—— (2005b), 'Hreyfingar ferðaþjónustu: Skipun þróunarverkefnis í ferðaþjónustu á Vestfjörðum', *Landabréfið* 21:1, 21–38.

—— (2007), *Emergent Tourism: An Actor-Network Approach to Tourism Economies* (PhD thesis, Department of Environmental, Social and Spatial Change, Roskilde University).

Latour, B. (1986), 'The Power of Association', in Law (ed.).

—— (2005), *Reassembling the Social: An Introduction to Actor-Network Theory* (Oxford: Oxford University Press).

Law, J. (1986), 'On the Methods of Long-distance Control: Vessels, Navigation and the Portuguese Route to India', in Law (ed.).

—— (ed.) (1986), *Power, Action and Belief: A New Sociology of Knowledge* (London, Boston & Henley: Routledge & Kegan Paul).

—— (2004), *After Method: Mess in Social Science Research* (London: Routledge).

Law, J. and Hassard, J. (eds) (1999), *Actor-Network Theory and After* (Oxford: Blackwell).

Law, J. and Mol, A. (2001), 'Situated Technoscience: An Inquiry into Spatialities', *Environment and Planning D: Society and Space* 19:5, 609–21.

Massey, D. (2005), *For Space* (London: Sage).

Mol, A. (1999), 'Ontological Politics. A Word and Some Questions' in Law and Hassard (eds).

Mol, A. and Law, J. (1994), 'Regions, Networks and Fluids: Anaemia and Social Topology', *Social Studies of Science* 24, 641–71.

Northern Periphery (2004), *The Interreg III Northern Periphery Programme* [website], http://www.northernperiphery.net/main-documents-g.asp, accessed 10 May 2006.

Shaw, G. and Williams, A.M. (2004), *Tourism and Tourism Spaces* (London: Sage).

Skaptadóttir, U.D. (1998), 'Coping with Marginalization. Localization in an Icelandic Fishing Community' in Aarsæther and Bærenholdt (eds).

Urry, J. (2003), *Global Complexity* (Cambridge: Polity).

Chapter 14

Outside In: Peripheral Cultural Industries and Global Markets

Dominic Power and Johan Jansson

Introduction

Culture is a pervasive fact of human life, and wherever we find ourselves we have become dynamic cultural creators. Whether at the centre of a world city or on the fringes of the northern periphery, culture flourishes. In recent times, the dynamics of cultural production have shifted somewhat (Power and Scott 2004): it is now apparent that culture is increasingly the product of commercial organizations and industries, and that those products flow relatively freely on world markets. The industrialization of culture and the globalization of cultural markets have meant that it is common for people in the far reaches of Northern Europe readily to consume exotic products such as Australian aboriginal music (Connell and Gibson 2004) or Hollywood films (Scott 2005). Equally, people in the rest of the world have increasing access to exotic oddities such as Nordic handicrafts, Scandinavian furniture or Swedish music recordings and performances (Power and Hallencreutz 2002).

The fluidity and ever increasing velocity with which cultural products move across borders present those interested in the cultural industries and those interested in the economic geographies of cultural production with an interesting set of challenges. In particular, we are left with the complex job of exploring and understanding interrelationships and dynamics between the place of production (as a locus for the production of a place-specific cultural product) and the spatialities of consumption and market that most often far exceed the physical boundaries of the place of production.

If cultural products are industrial products, then economic and industrial analysis has long drawn attention to the importance of the local milieu in a firm's successful creation of marketable and exportable cultural products. Indeed, studies of cultural industries have tried to understand why it is the case that certain *places* have been unusually productive crucibles for cultural production and innovation (Hesmondhalgh 1996; Scott 1999a; Scott 1999b; Scott 2000; DuNoyer 2002; Power and Scott 2004). In music, cities such as New York, Nashville and Liverpool have accounted for a disproportionate share of hits and artists; in fashion, London, Paris and Milan dominate the global scene; in film, firms based in Los Angeles and Bombay generate a volume of commercial success unrivalled by other industries. It is not surprising, then, that explanations of success in global cultural product markets have also tended to focus on the role of place in creating innovative and commercially successful exports. Scott suggests that such spatial agglomerations are not just 'spatial accumulations of physical capital, but also evolving pools of human skills and aptitudes'. These pools or communities of workers, he continues, are 'the preserve of accumulated traditions and conventionalized sensibilities … and they function as potent frameworks of cultural reproduction and arenas of socialization' (1999a, 1974). The message here, as in many modern agglomeration accounts, is that spatial clustering and agglomeration are important in creating and enacting different types of 'soft infrastructures' which, in turn, are crucial to commercial success. Competitiveness, even in highly global product markets, relies on a *local place-specific* capacity and milieu, where the sharing of 'conventions, common languages and rules for developing, communicating and interpreting knowledge' (Storper 1995, 206) makes for better and more competitive products.

For cultural producers in smaller countries and regions, spatial agglomerations can be hard to form or find. For those in the sparsely populated areas of Northern Europe, finding any firms/producers working in the same field can be nigh on impossible. It can be hard to find new employees and collaborators, source related services or even observe new trends and absorb new information and ways of doing things. Above all, there exists the basic problem that local markets are simply too small. If the cultural producers are interested in living off highly niched, avant-garde or sub-cultural products, such problems can be further exacerbated. Cultural producers in such places are in many ways forced to look outwards, and the opportunities offered by access to foreign markets, new publics, new ways of thinking and new people are vital. Indeed, when one talks to cultural producers in peripheral regions there is often a stronger desire to link into global circuits and chains than there is a desire to break free from them, or to focus solely on their place of origin.

The difficulty with both of the above accounts – the pull of the local agglomeration or of the extra-local – is that they are better adapted to explaining how 'static' or 'stable' products are produced or exported. The place of origin is seen essentially to imbue or imprint the product with place-specific markers that to a large extent explain the product's identity and marketable value (Molotch 1996; Molotch 2003).

These types of relatively static understandings of cultural offerings need to be questioned. It is true that most of the furniture produced in peripheral northern areas will not change material form when it is exported. However, in an economy and society increasingly dominated by social and commercial values, tradeability and a morphing of signs and symbols (Lash and Urry 1994; Baudrillard 1998), it is far from clear that we should think of furniture as simply a material product. The value and competitiveness of furniture from the Nordic countries has long been in the high levels of design, aesthetic and symbolism, of which individual pieces are markers (Fiell and Fiell 2002). Such immaterial aspects of cultural products are difficult to control and stabilize. Ultimately, it is the market and consumer appreciation of the products' meaning – place-specific or otherwise – that determines its success or lack thereof (Levitt 1975 [1960]).

Irrespective of whether the cultural producers in question are interested in commercializing elements of place-specific traditional culture or nature (for example, furniture made using traditional methods and materials) or whether they are residents of the northern periphery interested in radically new forms of cultural expression, it is their products' differentiation from competing substitutes that is most vital to commercial survival. Product differentiation and distinction, as Bourdieu points out, are highly embedded in the dynamics of the consumer milieu (Bourdieu 1984). Thus, we are faced with a situation where cultural products – irrespective of whether they are produced in a large centre or in a remote area – are imprinted with meanings and symbols of their origin that will undoubtedly be reinterpreted, relocated and resituated according to the specificities of the people and places they find their way to. Traditionally crafted Swedish furniture will thus be viewed very differently through the eyes of Parisian consumers than the eyes of fourth-generation Swedish-Americans in Minnesota. If we share such a view and are interested in explaining the dynamics of cultural product-based industries, then the question of origins – from the periphery or the centre? – becomes (though still important) slightly less interesting than the way in which they enter new markets, and how consumers receive them.

The rest of this chapter draws upon this idea of the importance of the dynamics of other places for cultural producers from the periphery. It shows that for Nordic furniture producers, a central route into other markets and into consumer consciousness is through trade fairs. Trade fairs are temporary events, but for the duration of their existence they function as dynamic sites of interpretation and symbolic construction. Their importance in smoothing products into diverse new markets and in the dynamic of immaterial production means that for producers in the Nordic periphery, trade fairs are far from isolated and distant events: they have lasting consequences for the organization of a firm's local setting and working, and should be seen as essential complements to the dynamics of local production.

The Road to Market: Furniture, Design and Trade Fairs

In empirical work that we have conducted over the last few years on the interior design and furniture industries,[1] we noticed that we were often split between the preoccupations of our theoretical backgrounds and the preoccupations of the Nordic firms we met. Our theoretical backgrounds pointed to the idea that if we wanted to understand the industrial and competitive dynamics of the cultural industries we

---

1   The empirical project upon which this chapter is based focused on collecting data during and inside trade fairs, and on research focusing on the industry and firms in day-to-day life, that is, not at fairs. Data collection during trade fairs involved several researchers and repeat visits in 2004–2006 to international trade fairs: principally Milan International Furniture Fair and Milan Design Weeks, and Stockholm International Furniture Fair. Fieldwork was informed by the ethnographic tradition and focused on participant observation, direct observation, analysis of documents produced within the industry and at the fair (for example, fair guides, press releases, catalogues), self-analysis, and conversations with visitors, exhibitors and organizers of different levels of formality (this involved everything from over 40 longer interviews to informal interviews, collective discussions and small talk). Since industry actors spend much of their year preparing for and following up on trade fairs, we conducted research with actors in their local settings (principally in Sweden). This involved close dialogue with over 50 designers and furniture industry actors in the Nordic countries. Dialogue and fieldwork were complemented by an extensive analysis of the wider literature, visual and textual narratives produced by top trade fairs.

were interested in then we should be out looking for vibrant local cultures, path dependencies and stable production centres, industrial districts, clusters, and so on. However, firms we met seemed less preoccupied with their day-to-day surroundings and local milieu than existing research had led us to think. For firms in industries as diverse as automobiles, music, computer games, boat-building, publishing, fashion, design and furniture, it appears that trade fairs, exhibitions and temporary events are major operational foci and the key way for cultural producers from the periphery to reach the market-place.

The central argument in this section is that trade fairs must not be perceived as isolated instances, but rather as events connected together in *global circuits* that are crucial to the organizational architecture, temporality and spaces of markets and the industries that serve them. In the design and furniture industry, global circuits based in part on participation in trade fairs form a vital architecture within which different industrial and market scales are mediated and connected. At the local level, these events are 'temporary clusters' (Maskell et al. 2004), but at a global or sectoral level their presence is much more permanent. They offer not only important publicity and marketing opportunities but also spaces where both knowledge creation and capital accumulation can occur.

The furniture (and interior design) industry is one where connections to other places through diverse intermediaries are becoming increasingly important. At the highest end of the furniture trade, designer-makers or traditional craftspeople control every aspect of design and production. In the Nordic countries these types of producers are distributed broadly, but primarily in rural areas. However, most of the furniture is supplied by mass-market furniture firms (especially large global retailers like IKEA and Conforama) or firms serving high-volume contract markets. These types of furniture firms rely on relatively disarticulatable production and design processes. For furniture firms with a home base in a high-cost Nordic country, many aspects of production are carried out extra-locally. Labour-intensive manufacturing processes are off-shored through flexible and diffuse supply and production chains, whilst strategic design, sales and marketing functions are kept in-house or out-sourced to firms in 'advanced' countries. The designer is seldom located proximate to production, and need never even visit the production facilities. Furniture production in Europe, for example, has thereby been forced by successive waves of painful industrial transformation to place design at the forefront of its strategic and innovatory efforts to remain competitive. This does not mean that localized centres of excellence do not characterize the furniture industry. In the Nordic region there exist a number of areas where middle to high-end furniture producers (predominantly small to medium-sized enterprises – SMEs), designers and craftspeople agglomerate: in particular, Möbelriket [The Furniture Kingdom] in Småland in Southern Sweden, and the Danish furniture cluster in central Jutland (Lorenzen 1998). While these agglomerations are not located in the most isolated areas of the Nordic region, they share with their compatriots in the farthest north a peripheral and distant relation to the largest markets for their products. Periphery is therefore a very relative concept: whether the furniture producer is located in Småland or in an isolated cabin within the Arctic Circle makes little difference if the buyers and consumers viewing and evaluating their products are in Paris or Minnesota.

In order to find connections to such actors and contexts, furniture designers and firms commonly use trade fairs. Trade fairs are central intermediary spaces and may be seen as microcosms of the industry they represent, with a multitude of buyers and sellers, service providers, partners, industry and regulatory bodies all gathered in one place to do business. During the few days of its life, the [trade fair] almost becomes the market (Rosson and Seringhaus 1995, 87).

For a short time, trade fairs are spaces where industry actors gather in one place. But these are spaces where visitors and exhibitors alike view their involvement as a crucial strategic investment for which they expect a return. The people and organizations we interviewed were categorical in their view that trade fairs were primarily economic events and platforms. They were seen to be central, in a multitude of different ways, to various business processes: but principally to sales and marketing.

Participants in trade fairs attend for a variety of reasons. Those who attend trade fairs do so, by and large, in order to reap some sort of economic or business benefit. Given the costs involved, this is no surprise. Our research shows that both exhibitors and visitors consciously use trade fairs as spaces within which to pursue particular business goals or aims. Some participants are clearly working to predetermined plans, with minutely-defined goals and objectives – such as meeting a set list of clients, or recruiting five new designers – whilst others take a less structured approach that is nonetheless based on the idea of getting value for money out of attendance.

We suggest that the trade fairs we examined exhibited six main types of activity space, each of them focusing on and dedicated to a tangible business or capital process. These six main spaces exist side by side for the duration of the trade fairs: spaces for sales; networking; marketing; knowledge flows/creation; recruitment; and institution building. The combination of a variety of functionalities, a crowded visual arena and a large number of visitors allows firms to pursue their interests intensely in connecting to new markets, people and knowledge. Against this background, participants appear to fluctuate between attention and distraction; between goal-oriented searching and getting caught up in the event or distracted by unexpected sights. It may be considered that this blend of structured information and chaos is what makes such events successful sites for making contacts and connections. They allow participants to pursue strategically planned and routinized forms of learning and interaction, alongside the unexpected information and impressions that the breakdown of normal routines and spaces brings. The sensory and experiential intensity of the spaces (within and beyond the fairgrounds) leads one to wonder if 'learning-by-playing' or 'learning-by-enjoying' are not central to the knowledge and inspiration gained by the participants. Much like bustling city streets or carnival seasons, fairs may be considered to be 'buzzing' (Storper and Venables 2002; Bathelt et al. 2004) with a density of 'thick and thin' knowledge 'in the air' (Gertler 2003).

Furniture and design trade fairs are spread across the year in such a way that it is possible to attend a number of them; they are mainly concentrated into two periods, one in the spring and one in the autumn. These periods coincide in time with when firms tend to launch their collections. In practice few firms, except for the largest, can afford the time or expense involved in attending all of them. Firms therefore plan their attendance/exhibition schedules on the basis of the fairs that are most important to their business aspirations.

For many Nordic firms, their domestic markets still represent a sizable portion of their sales. Thus, Danish firms will exhibit/attend Copenhagen as a matter of course, Swedish firms Stockholm, Finnish firms Helsinki, American firms High Point, and so on. The lower status and nationally focused trade fairs tend to display a variety of firms: from the very small to the very large. Over and above these local events is a hierarchy of global trade fairs. The top trade fairs tend to display the wares of only the very largest and most influential firms. However, all trade fairs have special spaces for younger designers and firms and in these areas many young (or small) Nordic furniture designers (and students) exhibit with a view to selling their products or getting new jobs.

Undoubtedly the largest and most prestigious international furniture and design trade fair is the annual Milan International Furniture Fair. The Milan fair has developed into a family of concurrent trade fairs covering not just furniture but also lighting, kitchens, bathrooms, office furniture and fittings, interior textiles and accessories: now commonly branded Milan Design Week. In total, almost 3,000 exhibitors were involved in the 2005 Milan Design Week, attracting 220,000 trade visitors and over 4,000 journalists. The presence of such a large and varied international press (everything from TV to specialist press) means that the fair attracts global media attention, which in turn further attracts firms to the fair.

Whilst Milan may operate as a global node in the trade fair circuit and the global furniture/design market-place, regional fairs also play an important role in both global and regional flows and markets. Worldwide, each region or large country tends to have one key furniture and design trade fair around which the regional market is organized and structured. In Scandinavia, the Stockholm Furniture Fair currently aims to perform this role. In 2006, the fair attracted over 30,000 trade visitors from 56 countries, and over 1,000 journalists. Following the lead of other such events, the fair is now part of a wider Stockholm Design Week, which attempts to spread the fair throughout the city, and indeed help to market Stockholm as a 'design capital'. The Stockholm Fair has become the pre-eminent trade fair within the Nordic region, functioning to serve not only imports and exports to and from the Swedish furniture and design market, but also the Nordic region in general.

The costs involved in participating in trade fairs can be very high. Simply attending a trade fair not only involves admission charges but also travel and personnel costs. In extreme cases, like the Milan Design Week, hotels raise their room rates to three or four times the regular level. Likewise other business facilities, such as catering and conference rooms, are expensive and hard to find.

For firms that exhibit, the costs involved can be extremely high. Costs are incurred before, during and after the fair, when firms follow up on deals made and contacts brokered at the event. Attending a major event such as Milan represents a major investment. The smaller Nordic firms we interviewed reported an average spending of €50,000 for the stand alone. Of that, €30,000 was spent on hiring the stand and another €20,000 on the design of the stand. This figure does not include costs relating to manufacturing prototypes, personnel, travel and hotels. For larger and more prestigious firms, it is not uncommon to spend up to €200,000 on stand design alone. Additionally, very large sums are spent at off-site events and formal and informal festivities. The level of these costs goes a long way in explaining why only the largest firms exhibit. Such costs also represent a significant entry barrier for firms wishing to expand into new markets or go global.

The costs and the expectations attendees have concerning both the products and the exhibition stands themselves mean that preparing for a major show such as Milan or Stockholm may occupy almost the entire year leading up to the event. Indeed, the stand itself is often a major design product. Stands at large fairs in market segments such as kitchens, bathrooms and office environments are large and expensive, looking more like architect-designed buildings than market stalls. In large fair grounds, the placing of the stand itself is an important factor in attracting passing trade and maintaining the right profile. Stands located on the main corridors and facing entrances are highly prized; so, too, are stands neighbouring particularly high-status competitors. Neighbourhood effects and agglomeration effects seem to play an important role, even in a temporary setting.

For the exhibiting Nordic firms we interviewed, the business year is essentially structured around a series of deadlines imposed by the global circuit of trade fairs in which they are involved. Principally, firms will attempt to launch new products or ranges at the largest/most prestigious trade fair they regularly attend. Due to the fact that the furniture produced by the firms we interviewed is most often sold in large quantities to contractors, retailers or wholesalers, in reality firms only need high-quality prototypes or demonstration models for display. Some firms we interviewed stated that they had worked until the very last minute getting these pieces ready for the show. In general, firms reported that they had new pieces and ranges ready for the show between one and two months in advance. Designing, prototyping and testing a new product can take many months. Following this will be a period of (re)organizing production facilities or new sub-contractors in readiness for production. It has also become common practice for the more design-based leading firms to display showcase or concept pieces that may grab media attention and raise their reputation for innovation – just like automobile manufacturers. Such pieces do not place the same demands on firms as ready-for-production pieces, but they nonetheless involve significant design input and often costly model and prototype manufacture. This type of product development and prototyping is most usually geared towards a launch at a specific fair. Thus, exhibitors often develop and present both a ready-for-sale range and a concept range, as well as carefully designed and assembled stands and events.

During the period directly leading up to the fair, an average of one month is spent preparing press materials, catalogues, and so on, and in contacting the media and (actual and potential) clients for meetings at the trade fair and related activities. Magazine and trade advertising for months ahead is often directed towards delivering the message that the firm will be at a particular fair at a particular stand and/or off-site event. Such work is not uncommonly subcontracted to agents and promoters specializing in specific trade fairs or particular segments. Large firms may employ a full-time team responsible for trade fair participation. Preparation may also be required, not just for the trade fair event itself but for the parties, extra events, special shows and private viewings that are now expected of the larger firms: there is considerable pressure for these to be both fun and well-designed. After the trade fair, a range of follow-up activities must be conducted: not just following up on clients but also on media coverage, competitor analysis and further interviews with potential employees/sub-contractors, and so on. Attendees interviewed said that this would occupy them for between three and five weeks after the fair.

The more firms pursue monopolistic competition (Chamberlin 1933) as a strategy, or produce stylized products/services that appeal to very particular/specialized tastes, the more likely it is that they will not find sufficient demand locally and will therefore need actively to seek out connections. With the exception of those firms that own their own distribution and retail channels (for example, IKEA), large and small firms must actively seek out extra-local customers (mainly retailers, but also architects, interior designers and contractors). Trade fairs provide unique spaces where geographically dispersed supply and demand can meet. As such, these global fair circuits are central components in the architecture within which different industrial and market scales are mediated and connected. In this way, they are less 'temporary clusters' than they are '*cyclical clusters*': they are spaces that are timed and arranged in such a way that cultural product markets and innovations can be enacted, reproduced and continuously renewed over time.

Conclusion

Culture is not entirely enacted and created in the local place. In many cultural product industries, such as furniture, connections outside the local area are vital, not only for making business connections but for the 'construction' and negotiation of their products' meanings and thereby their values. Trade fairs are important in this respect. However, they are far from isolated events and have an effect on a firm's entire year and context. Thus, such events need to be regarded not as add-ons to the local creative production place but as part of a set of global circuits that link together and mediate between places and spaces constitutive of the commodity chain. Thus, we introduce the idea of global circuits and '*cyclical clusters*', which are timed and arranged in such a way that cultural product markets and innovations can be enacted, reproduced and continuously renewed over time.

The structures and dynamics reported in this chapter have implications for those interested in the development of peripheral cultural industries. Throughout our research, no one ever questioned the idea that excellent musical products, or pieces of furniture or design, were produced outside the orbits of large cities or in the farthest northern peripheries. Indeed, in the cultural product market, unique origins or place-based associations most often help to differentiate the products themselves in positive ways (Levitt 1975 [1960]; Levitt 1980; Levitt 1981; Molotch 1996; Molotch 2003). Rather, it is the structure of certain large markets, media channels, gatekeepers, or access to distributors, retailers and consumer tastes that is most often singled out as the reason why products can have such a difficult road to market. Cultural producers on the periphery are often aware of these hurdles and frequently organize their year and efforts around global circuits and the dictates and rhythms of foreign markets.

These findings both contradict and complement explanations concerning success in the cultural industry centred on a dynamic creative milieu: that it is largely due to the competitive, sometimes even artistic, qualities of the things produced in certain places that market success is secured. Indeed, a growing literature on the complexities and inequalities inherent in global commodity chains indicates that no matter how good a product may be, it is the conditions of access to markets and product reception that are crucial in dictating a product's success (Gereffi 1994; Gereffi and Korzeniewicz 1994; Negus 1999). The manufacture or origin of cultural products should therefore be considered as only a first step in the commodification and commercialization processes that determine their success. This implies that power over the commercial success of export-oriented products is also 'located' outside the local production milieu. Enacting and supporting peripheral cultural industries does not involve local processes alone: it involves looking further afield and linking up with seemingly far-off places and processes. We must therefore attempt to develop accounts of peripheral cultural industries, and the policies directed towards supporting them, that recognize the fact that whilst enacting local cultures and encouraging mobility within the periphery is important to creative cultures and their economic potential, in a globalized world it can only ever be part of a much more complex narrative.

In conclusion, this chapter strongly echoes the claims made by Jóhannesson and Bærenholdt in Chapter 13, that some sort of relational approach emphasizing connections through networks (be these networks of actors or commodity and production chains) is necessary to explain and understand activities in remote places.

It becomes meaningless to talk of local versus global processes as in much of the global-local literature; instead we should think in terms of networks of agents (such as individuals, institutions or objects) acting across various distances and through diverse intermediaries (Dicken et al. 2001, 95).

Thus, mobility and place cannot be thought apart from each other and we must continue to direct our efforts to understanding the complexities inherent in manifestations of space-producing practices: practices which are seldom clearly local, global or extra-local.

References

Bathelt, H., Malmberg, A. and Maskell, P. (2004), 'Clusters and Knowledge, Local Buzz, Global Pipelines and the Process of Knowledge Creation', *Progress in Human Geography* 28:1, 31–56.

Baudrillard, J. (1998), *The Consumer Society, Myths and Structures* (London: Sage Publications).

Bourdieu, P. (1984), *Distinction: A Social Critique of the Judgement of Taste* (London: Routledge & Kegan Paul).

Chamberlin, E. (1933), *The Theory of Monopoly Competition* (Cambridge MA: Harvard University Press).

Connell, J. and Gibson, C. (2004), 'World Music, Deterritorialising Place and Identity', *Progress in Human Geography* 28:3, 342–61.

Dicken, P. et al. (2001), 'Chains and Networks, Territories and Scales, Towards a Relational Framework for Analyzing the Global Economy', *Global Networks* 1:2, 89–112.

DuNoyer, P. (2002), *Liverpool, Wondrous Place, Music from Cavern to Cream* (London: Virgin Books).

Fiell, C. and Fiell, P. (2002), *Scandinavian Design* (Köln: Taschen).

Gereffi, G. (1994), 'The Organization of Buyer-driven Global Commodity Chains, how US Retailers shape Overseas Production Networks', *Commodity Chains and Global Capitalism*, G. Gereffi and M. Korzeniewicz (Westport, CN: Praeger).

Gereffi, G. and Korzeniewicz, M. (eds) (1994), *Commodity Chains and Global Capitalism* (London: Praeger).

Gertler, M. (2003), 'Tacit Knowledge and the Economic Geography of Context or The Undefinable Tacitness of Being (There)', *Journal of Economic Geography* 3:75–100.

Hesmondhalgh, D. (1996), 'Flexibility, Post-Fordism and the Music Industries', *Media, Culture and Society* 18, 469–88.

Lash, S. and Urry, J. (1994), *Economies of Signs and Space* (London: Sage).

Levitt, T. (1975 [1960]), 'Marketing Myopia', *Harvard Business Review* 53:5, 26–42.

—— (1980), 'Marketing Success through Differentiation – of Anything', *Harvard Business Review* Vol. 58:1, 83–92.

—— (1981), 'Marketing Intangible Products and Product Intangibles', *Harvard Business Review* 59(3), 94–103.

Lorenzen, M. (ed.) (1998), *Specialisation and Localised Learning, Six Studies on the European Furniture Industry* (Copenhagen: Copenhagen Business School Press).

Maskell, P. et al. (2004), 'Temporary Clusters and Knowledge Creation: The Effects of International Trade Fairs, Conventions and Other Professional Gatherings', *Spaces* 04.

Molotch, H. (1996), 'LA as Product: How Design Works in a Regional Economy', in Scott and Soja (eds).

—— (2003), *Where Stuff Comes From: How Toasters, Toilets, Cars, Computers and Many Other Things Come to Be as They Are* (London: Routledge).

Negus, K. (1999), *Music Genres and Corporate Cultures* (London: Routledge).

Power, D. and Hallencreutz, D. (2002), 'Profiting from Creativity? The Music Industry in Stockholm, Sweden and Kingston, Jamaica', *Environment and Planning A* 34:10, 1833–54.

Power, D. and Scott, A. (eds) (2004), *Cultural Industries and the Production of Culture* (London: Routledge).

Rosson, P. and Seringhaus, F. (1995), 'Visitor and Exhibitor Interaction at Industrial Trade Fairs', *Journal of Business Research* 32, 81–90.

Scott, A. (1999a), 'The US Recorded Music Industry: On the Relations between Organization, Location, and Creativity in the Cultural Economy', *Environment and Planning A* 31, 1965–84.

—— (1999b), 'The Cultural Economy, Geography and the Creative Field', *Media, Culture and Society* 21, 807–17.

—— (2000), *The Cultural Economy of Cities: Essays on the Geography of Image-producing Industries* (London: Sage).

—— (2005), *On Hollywood* (Princeton NJ: Princeton University Press).

Scott, A. and Soja, E. (eds) (1990), *The City, Los Angeles and Urban Theory at the End of the Twentieth Century* (Berkeley: University of California Press).

Storper, M. (1995), 'The Resurgence of Regional Economies, Ten Years Later, the Region as a Nexus of Untraded Interdependencies', *European Urban and Regional Studies* 2, 191–221.

Storper, M. and Venables, A. (2002), *Buzz, the Economic Force of the City* (DRUID Summer Conference on 'Industrial Dynamics of the New and Old Economy – who is embracing whom?', Copenhagen/Elsinore).

Chapter 15

Place and Transport: Particularities of a Town and its Mobilities

Brynhild Granås and Torill Nyseth

Introduction

Until recently, the town of Narvik in Northern Norway was known for its prosperity. Ranging from its cultural life and the athletes of the town to its welfare services, Narvik was 'top ranked' in the region. Similarly, the inhabitants were noted for the pride, and sometimes even haughtiness, they displayed as women and men of this town. Over the last century, much of the economic status and stability of Narvik relied on iron ore transport. Transport of iron ore is still a manifest part of life in Narvik. Nevertheless, year by year since the 1960s, it has demanded less space and many fewer hands at work. Today the activity is no longer a main source of nurture for the town, nor its pride or its economy.

After a while, the recession that set in discreetly from the 1960s onwards precipitated a search for ever new ways of making a living. Finally, at the beginning of the twenty-first century, a dawning optimism has been discovered and a diversity of businesses now comprises the local economy. Strong, but nevertheless diffuse antagonism between local leaders, who display their disagreement in public, is also part of today's picture, as is a growing new version of the transport industry.

The town of Narvik exemplifies a particular instance of an integral relationship between mobility and place. The following analysis will focus on the social production of the place and its mobilities, with particular attention paid to the transport industry. This industry is constituted by the movement of raw materials and goods, train wagons, boats and trailer lorries, as well as corporeal movement. Social, physical-material and symbolic aspects of mobile transport-related practices will be explored. Part of our argument is that the mobilities involved have constituted the place from the beginning and articulated distinct place characteristics at specific moments of its history. In the course of the chapter we suggest that the concurrence of optimism, antagonism and growth in new transport industrial activity that is now to be found expresses changes, continuities and particularities in the role of mobilities within the production of place. This chapter is based on fieldwork undertaken in the town of Narvik in 2006.[1] A total of 36 people were interviewed individually and in group interview sessions.

1   This research is part of the Nordregio-funded project 'Place Reinvention in the North. Dynamics and Governance Perspectives' (Nyseth and Granås 2007).

The Social Place and its Mobilities

Both 'place' and 'mobility' are concepts that refer in everyday language to phenomena which are so 'normal' and 'universal' that they may be characterized as 'uninteresting' (Cresswell 2006). Still, as Tim Creswell (2006, 22) states, 'It is inconceivable to think of societies anywhere without either (place or mobility), and yet any particular way we have of thinking about them is…socially produced'.

Our approach relates to practice theory (Massey 1994, 2005; Simonsen 2001) and perceives places as produced within social interaction; the place is perceived as progressive, extrovert and time-specifically articulated (Massey 1994); it is always becoming, practised within relations that often reach far beyond the physical borders of place.

> Instead…of thinking of places as areas with boundaries around, they can be imagined as articulated moments in networks of social relations and understandings, but where a larger proportion of those relations, experiences and understandings are constructed on a far larger scale than what we happen to define for that moment as the place itself. (Massey 1994, 154)

The descriptions that follow from the town of Narvik will account for place in such progressive and extrovert terms, with special attention paid to the relational practices of the transport industry. Furthermore, the descriptions emphasize physical-material features of place: the social and negotiating character of place involves not only people, but also the 'non-human' (Massey 2005, 140; see also Chapter 17 by Benediktsson), for example, natural qualities, land structures and physical constructions. Physical-material interventions of transport practices are severe: the industry is space-consuming and leaves a vast physical footprint, inescapable to the eye and unavoidable for any town planner. This aspect is important when looking at transport and place. Our account shows how this has been made especially topical with regard to land use negotiations.

In addition, place is a matter of meaning and subjectivity that is constituted through participation and involvement (Simonsen 2001). This chapter emphasizes such considerations of subjectivity, whilst also stressing experiences and meanings of transport-related practices and mobilities. At this point, we relate our argument to Cresswell, who states that 'The movements of people (and things) all over the world and at all scales are, after all, full of meaning' (2006, 2). He makes an analytical distinction where 'movement' refers to something apparently natural, devoid of meaning, and 'mobility' points to movement as imbued with meaning and power (2006, 3). Our description tells of the relationship between the place and transport-related mobilities, and accounts for meaning, power and politics produced by and producing these mobilities. Our approach to the theme of power is also inspired by Foucault's exhortation to make power *relations*, not power itself, the object of analysis (Foucault 1982, 219). We perceive this emphasis to sit well with an understanding of place as socially relational, characterized by negotiation.

Further, Cresswell claims that 'If movement is the dynamic equivalent of location, then mobility is the dynamic equivalent of place' (2006, 3). Even though we agree that place is socially produced and imbued with meaning, this chapter, in line with

Massey (1994; 2005; 2007), argues against a framing of place as static (cf. Bauman 2000; Cresswell 2006). Furthermore, mobility and place should not be perceived as dichotomies. Rather, this chapter explains how the social production of place includes the practice, experience and meaning of mobilities.

Finally, our ambition has been not to present a history of the place in such a way that we intend to re-establish continuity, or show how the past is ever present and 'secretly animates our contemporaries' (Foucault 1971). Even though we tell of the relationship between the transport sector and the town of Narvik, we do not claim that this is the only story from Narvik that could be told: that Narvik 'is a place of transport', for example, or that the continued transport now to be found is a 'necessary' outcome of history. Rather, this account from Narvik follows Massey, when she states that it is '...precisely...throwntogetherness, the unavoidable challenge of negotiating a here-and-now' (2005, 140) that makes up the particularities of place.

The Creation of a Railway – and a Town

The town of Narvik sprang up around the end of the nineteenth century, located in the middle of Northern Norway on a narrow strip of land between the deep Ofoten Fjord and the steep mountains behind. This land was finally chosen as the transhipment port when Swedish interests required a transport line to start industrial production of iron ore in the rich mining areas around Kiruna in Sweden. The railway construction through the mountains was completed in 1902; the Ofoten Railway was opened and the first iron ore wagons arrived by train at Narvik harbour. The same year, the town was formally established. The iron ore that flowed in by rail and out via Narvik harbour on iron ore ships entailed the formation of a town on this small piece of land. Narvik came into being, developing from a handful of small-scale farmers' families into a town of several thousands within less than ten years (Aas 2001), as the modernity of the railway and industry so suddenly and intensively arrived. Today, the town has 14,000 inhabitants.[2]

The Significance of Iron Ore-related Mobilities

Right from the start, transport practices entailed the formation of power relations in the town. Firstly, Narvik was a concern of two nation-states, initially as part of the Swedish-Norwegian Union and after 1905 within subsequently independent Norway:[3] the extensive railway construction was co-financed by the Norwegian and Swedish national authorities. The main employers in the new town were the Swedish and soon nationally-owned mining company, Luossavaara-Kiirunavaara Aktiebolag

---

2   Narvik municipality comprises 18,500 inhabitants.

3   When the Swedish King Oscar II visited the town in 1903, he arrived by train and was welcomed at the train station some kilometres from the harbour. In 1907, two years after the union was dissolved, the Norwegian King Haakon VII arrived by boat and received his welcome at the harbour (Aas 2001). 'This was the Norwegian way of arriving in town', says the historian Steinar Aas.

(LKAB) and the Norwegian State Railways (NSB). NSB owned the railway line and was responsible for the transport. NSB and LKAB became dominant actors within the town of Narvik throughout the twentieth century, as well as representing the two national authorities involved.

Secondly, the municipal institution of Narvik became economically strong, a position that relied very strongly on a lucrative tax agreement with LKAB from 1913 onwards. This made Narvik the richest municipality in Northern Norway for decades. Finally, the working class, dominated by NSB and LKAB employees, soon co-ordinated its efforts on the local political scene and the social democratic Labour Party has been in power ever since 1913 (Aas 2001).

The mobility involved was both a product of power and produced power in ways such as those described above (cf. Cresswell 2006). In addition, the roles of NSB and LKAB, especially, exemplify very well the extrovert and relational characteristics of place and the power geometries involved (Massey 1994; 2005) between the national centres and this small periphery town. LKAB, in particular, and by extension the Swedish national authorities, held a very strong position in place negotiations: on several occasions, LKAB threatened to make use of alternative iron ore transport routes (Svendsen 2002). Throughout the twentieth century, moreover, the municipality, inhabitants, LKAB and NSB shared a destiny of recessions and revivals in the global iron ore market, together with the mining towns in the Swedish mountains.

In line with the arguments of Creswell (2006), it may be stated that the railway line was certainly not just a 'neutral' physical line, stretching from Kiruna to Narvik, on which trains were 'impartially' placed to move iron ore. Their movements were imbued with power and politics, as described above, as well as meanings of other kinds.

Firstly, the train represented 'the arrival of modernity' within a region that otherwise lacked railway connections and was considered a remote periphery of Norway. It prepared for mobilities that allowed Narvik inhabitants to experience their town as more modern and central, in stark contrast to its surrounding districts. Narvik was suddenly connected to Stockholm and the European continent. Inhabitants tell stories of how they were able to secure relations with Swedish society and feel the breath of the outside world as the train glided through town on a daily basis. 'During the 1960s we were rather proud: we had a train connection to Stockholm…', says one informant. The first decades also offered the possibility for young Narvik men to sign on as crew on the iron ore ships. Stories from such trips are still heard in Narvik. Even though most people used not to travel to the European Continent or to ports abroad, our interviews suggest that the imagination of this mobility (Cresswell 2006, 3) is what inspired the inhabitants' feelings of being important.

Secondly, the years of railway construction gave rise to a technical industrial culture, as well as endlessly vivid narratives about the 'origins' of the town. The construction was an opening shot for the strong technical competence (closely pursued by the technical management of the iron ore transport itself), of which we can still see traces today in manifestations such as technical businesses and educational institutions. Narratives of the construction years have been celebrated at the very popular local Winter Festival Week (*Vinterfestuka*) since the 1950s.

Thirdly, the lack of extensive or good-quality road networks in the region caused the train to become a strong transport actor. Goods were transported on the railway and by boat, and the distribution of goods characterized town life during the initial decades. The railway goods traffic lasted until the 1960s, when the Nordland Line was completed, extending the Norwegian national railway network to Bodø, 300 kilometres south of Narvik. Northern railway goods transportation was thereafter monopolized by the NSB Nordland Line. Nevertheless, ever since then, most of Northern Norway has been left without any railway connection other than the Ofoten Railway.

So, from early on, the iron ore transport provided for mobilities that empowered Narvik inhabitants in several ways. One was the potential for travel and experiences that gave life to narratives and 'imagined geographies' of a wider physical-geographical horizon, framing Narvik as a 'modern town'. Another was the construction of the railway line and the push that this phase provided in the development of technical competence in the town, as well as the strong place narratives and symbols of 'the historical origin of the town'. Thirdly, the line affected a wider transport industrial development that had a subsequent impact on the progress of today's new transport industry, described later in this chapter.

The Physical-material Place and Transport

Altogether, the above accounts portray socially constructed mobilities as an integral part of the production of place. Furthermore, this process also has its physical-material aspects. 'Places pose in particular form the question of our living together', states Massey (2005, 151), and negotiations involved take place '…within and between human and nonhuman' (Massey 2005, 140). We shall now focus on some 'nonhuman' aspects relating to the transport industry, its mobilities and the production of place.

One aspect is the natural quality of the land structure: Narvik's deep and sheltered harbour and its proximity to the Swedish mountains were qualities that attracted external investment to the place from the start, thereby enmeshing the land into far-reaching economical and political processes. But Narvik is contained in a tight spot: the town is a clearly disparate physical landscape, with limited land available. Another 'non-human' aspect is the physical-material and land-demanding character of the transport industry: a massive physical infrastructure is an essential part of this business. Similarly, the meeting of railway and boats (and later also trailer lorries) resulted in further space requirements, especially for reloading, but also partly for storage.

Nature and the limited land available led to a continuous contest for space, especially as the transport sector was such a space-consuming one. At an early stage, the land use regimes of the iron ore transport actors formed the physical contours of Narvik that are visible today (see Figure 15.1). In around 1900, the Norwegian national authorities, eager to secure Norwegian interests, took control of approximately 70 per cent of the town area, including Narvik harbour. LKAB also took control of large areas, including the train station area (Aas 2001).

Figure 15.1   Aerial photo of Narvik

Source: photo by Narvik Port Authority.

Narvik became the 'split' town that we can still see today. The railway line stretched from the train station in the east to the main harbour in the west, cutting the town site into two parts. In the middle is the town centre. The area that stretches from the centre down to Narvik harbour, 'the Triangle', was reserved for the activities of LKAB[4] and NSB throughout the twentieth century.[5]

This exemplifies how the movement of iron ore was imbued with external relational power, which impacted in decisive ways on the physical-material construction of the town. Right from the start, the new settlers complained that the interests of the iron ore transport business were being given priority at the expense of any concern for a well-functioning town life. One particular concern was that industrial activity within the Triangle prevented the emergence of a true coastal town, with a more appealing waterfront setting. Other complaints concerned a lack of access on the part of other industries to the harbour facilities (Aas 2001). But the physical structures that were laid down 100 years ago were to become stable aspects of twentieth-century Narvik. By the beginning of the following century, though, the taken-for-granted and stable character of this way of using the land had been turned upside down. As we shall argue later, today's negotiations concerning the land use of the Triangle articulate new meanings of mobilities in Narvik.

---

4   LKAB established a long-term agreement with the Norwegian state concerning beneficial rights within this area.

5   The historian Steinar Aas (2001) compares the railway line with a river that cuts through the land, and the industrial area towards the harbour with a river delta.

Palmy Days and the Long-drawn-out Recession

The recession that started in the 1960s and escalated in the 1990s was a long-drawn-out process that eroded the previous stability. At their peak, right after the war, LKAB and NSB employed 1,600 people in Narvik: 1,000 at LKAB and 600 at NSB (Svendsen 2002). Noises, smells and dust surrounded the iron ore activity on the railway line and within the Triangle. In retrospect, inhabitants speak of this as the heartbeat and breath of Narvik; the post-war period is perceived as the palmy days, of which the 'imagined mobilities' described earlier form a part. The labour class of these years is referred to as 'well fed', almost 'bourgeoisie'. Jokes are made about how arrogant townspeople may say to non-locals, 'Move yourself, I want to spit!'

From the 1960s onwards, LKAB rationalized its activities year by year. However, several other industrial projects buoyed people's spirits. From the remarks made by inhabitants, it is hard to identify any major event or specific year that turned into any sort of 'crisis'. However, one significant dispute is identifiable from the 1990s, when LKAB forced NSB to hand over the right to transport iron ore, with the resulting loss of most NSB jobs in Narvik. In addition, the municipal tax income from LKAB was reduced severely from the 1960s onwards (Svendsen 2002).

In 2007, there are only a dozen jobs remaining in Narvik within the old, but now restructured NSB Corporation. Similarly, around 200 people work for LKAB. The recession brought about a more heterogeneous local economic pattern. The dominant positions of NSB and LKAB were dissolved. Similarly, the vigour of the municipal institution was weakened, as was the strong position of the Labour Party, which in 2007 leads a political minority coalition. As the mobility relating to iron ore changed, its interlinked power relations fragmented.

A New Version of Transport Arises

By the end of the twentieth century, a new version of the transport industry had grown up. In what follows we shall describe how that happened. Our account shows that the outcome was not one of historical necessity, but rather one of concurrent externally related events, illustrating again how place is a particularity of linkages to an 'outside', '…which is therefore itself part of what constitutes the place' (Massey 1994, 155). The new transport industry developed in line with other trade branches within the town and was not given any obvious priority or attention. One implication of this line of development has been that the transport industry today participates in negotiations of place without enjoying the dominance once exercised by LKAB especially, but also by NSB, and we shall return to this in due course.

During the 1990s, when Narvik inhabitants were investigating new businesses, the transport industry – rather surprisingly, we are told – rose in slightly new clothing as the transport of goods was restarted on the railway. This transport had disappeared during the 1960s. In the 1990s, however, NSB saw a politically and economically profitable future in re-establishing the transport of goods on the Ofoten Railway. The EU had liberalized the railway sector during the 1990s[6] and placed an increased profit

---

6   Norway was bound by these new regulations through the EEA agreement.

demand on railway businesses. The economic rationale of NSB's monopolization of goods transportation on the Nordland Line which only reached Bodø, south of Narvik, was questioned. At the same time, handing over the iron ore transport to LKAB had placed NSB under pressure to compensate for the losses endured by the Narvik community. On these grounds, NSB now decided to embark on a joint project with Swedish Railways (SJ) to create an Arctic Rail Express (ARE), transporting goods between the north and south of Scandinavia, through Sweden and on to Narvik. SJ was eager to be involved in this project, in order to position itself in the approaching political process concerned with a restructuring of the Swedish railway system. 'It was a trump card in a high-level political play to prove that "the iron ore line" had other potential', said one Narvik informant, central within the former NSB system.

At the same time, Narvik provided a large pool of railway expertise. Some of these were people in the midst of their working careers, with the insider's understanding of the dynamics within NSB and the restructuring process the company was now going through. This competence and insight was made use of in a fight to ensure new goods-related railway activities in the town. Even though the goods transport activities from the first half of the twentieth century had been hampered by the monopoly of the Nordland Line during the 1960s, Narvik still had a considerable freight and distribution business, though not on the railway. The early goods transport history and the location of the town was prepared for this. Solid, but small wholesale and distribution companies encompassed competence that provided the strength to promote new activities in Narvik within the new railway regime. In addition, the robust physical construction of rails, double tracks and stations was lying there quietly in preparation for new usages in the future. The recession in the global iron ore market during the 1980s had become a severe threat to LKAB and their northern iron ore production. Fortunately for the Ofoten Railway, however, LKAB had overcome this dramatic crisis and the line was kept from moss and disintegration by continued iron ore transport.

All these elements prepared for the start-up of the goods transport project Arctic Rail Express (ARE). The first ARE train headed off towards the Swedish border in 1993 and was a success from the start. 'ARE has been a catalyst for the transport industrial efforts that are here now…', says one wholesaler company leader. Following ARE, branches of several national wholesale and distribution companies have been established in Narvik. Similarly, local companies have started to flourish. Recently, efforts have also been made to develop railway goods transport routes within the Barents Region and to extend ARE activity towards the European Continent.

The physical-material manifestations of the goods transport industry are to be found at the Fagernes Terminal in the southern outer harbour. A railway line stretches through the Triangle out to the quayside, where trailer lorries and boats are served by huge container cranes. The area is distant from the main parts of the Triangle and cannot be seen from the town centre. Business is made up of a range of companies, with a multitude of external and local owners. The activity is characterized by diversity, in contrast to the former dominance exerted by LKAB and NSB. Negotiations between the many parties involved are marked by tensions regarding land use: the simple addition of a continental ARE train could expand the scope of Fagernes Terminal.

N.E.W.: The suspense of a mega-project

In addition to all this there exists the Northern East West Corridor project (N.E.W.). At a public meeting in Tromsø, in 2006, administrators involved in the promotion of Norway's new High North policy commented that 'In Narvik they plan to become the new Singapore', and all those present laughed. Among industrial transport actors in Narvik, though, this is no joke. The basic idea is to transport goods by rail from China to Narvik, and then reload them onto ships bound for the USA. The developments described above have, together with high-level political changes, brought up this idea: the fall of the Iron Curtain, the increase in world trade, a consequent expansion of the total volume of freight, and similarly capacity problems and present-day security concerns (Transportutvikling Ltd. 2004).

The N.E.W. idea was first launched locally, but after a while external actors, such as the International Union of Railways (UIC) and Nordland County, have taken on leading roles. The complexity of N.E.W. seems very great, incorporating many nation-states,[7] a great span of institutions and actors, market issues and financial concerns. The planning and co-ordination challenges are similarly extraordinary.

The land available within the town today will not accommodate even the most modest volume prognosis for a China-USA freight corridor. Central actors describe plans to extend the railway through a mountain to a small village where the 'new Singapore' is to manifest itself. But, similarly significant, this phase of escalation will depend on considerable physical elbow-room with regard to available land. Even though N.E.W. is still at a planning stage, and the realization of it is inconclusive, this mega-project of mobility adds fire to the continuous negotiations for space in Narvik town.

Recapturing the Triangle – Negotiating New Ground Rules

The new transport industry, including the suspense surrounding the N.E.W. mega-project, has intensified requests for available land in the town of Narvik. To clarify which area within the town has become particularly involved in such negotiations, we shall now return to the changing positions of LKAB and NSB, since the reduced roles of these two companies also affected their land use: LKAB had leased land within the Triangle from NSB, but by the end of the 1980s LKAB withdrew from considerable sections of this area and expanded towards the western outer harbour. Parts of the Triangle, still owned by NSB, were laid to waste. In 1987, the municipal chief officer claimed that '…it was "fatal" for the development of Narvik that the area was not in use' (Svendsen 2002, 538). After lengthy negotiations, the Triangle was handed over from NSB to the municipality. A broad mobilization process was launched to contribute ideas as to how to develop the 'rescued' land. Many inhabitants suggested a makeover of the whole Triangle for parks and recreation. Despite this, the most manifest outcome to date has been the construction of a standard shopping centre and a bus terminal.

---

7   The railway transport line involves China, Kazakhstan, Russia, Finland, Sweden and Norway. The sea lane mainly relates to the USA, but Canada and Iceland are also involved.

The striking antagonism that we have hinted at earlier appeared in a recent, harsh public debate relating to the Triangle. The main area of disagreement seems to concern the actual organization of land planning within the area. Our interviews tell clearly of personal mistrust among alliances of people involved in public, semi-public and private development work, and the level of tension implies that much is at stake for interested parties. A development plan for the Triangle was approved politically in 1999, preparing for an industrially strategic use of the land, combined with culture, sports, parks and residential areas. However, several actors question its status. We asked leading local actors to comment on whether there is an underlying plan to hold back the area for industrial transport purposes again. Most actors involved were evasive and indicated that interested parties have agreed to follow the official development plan.

Some unambiguous statements, though, puncture the announced agreement and show that there are tensions, and that perceptions differ regarding the future of the Triangle. Firstly, our transport industry informants still consider the Triangle to be 'the most strategic area' of the town and regret that the area was abandoned some years ago. Secondly, transport actors are now making new claims within the Triangle. Thirdly, one of the central stakeholders of the N.E.W. project refers to the Triangle as '...a cadaver...(where)...a lot of people are trying to position themselves to get their share...'. Fourthly, a business leader in another sector complains that 'What could have become a seafront that could provide the town with an identity towards the sea has now been stopped because of N.E.W.'. Finally, follow-up interviews have also confirmed that the antagonism that marks town-life today is displayed within a continuing contest for the Triangle area.

New Meanings of Mobility Within the Production of Place

Descriptions of the growth and dynamics of the new transport industry, and the conflicts surrounding the reclaimed Triangle area, demonstrate a situation where the ground rules for negotiating place are undergoing transformation. The 'release' of the Triangle has encapsulated a more general situation in the town, where the space has been expanded for a variety of local interests to make claims in place negotiations. The transport industry no longer enjoys a dominant position within such negotiations. The mobilities of goods transport do not produce power (cf. Cresswell 2006) in the same way that iron ore transport once did: the business actors are numerous, with no obviously dominant ones, and the municipal income from the transport sector is moderate, on a level with other businesses. Similarly, the number of jobs involved is still limited: employees are spread over many different companies and the political situation in the town is marked by fragmentation and a weak mandate of political leadership.

The power and politics of mobilities have changed with the new transport industry. The same goes for other meanings relating to the mobilities involved: we were told that the inhabitants certainly paid attention to the ARE train when it started. But interviews tell us that, in general, goods transportation and distribution does not receive any particular attention locally. Recent history seems to moderate any high hopes among inhabitants that any single economic activity will secure future welfare. So far, the main focus of attention paid by the inhabitants to the new transport industry concerns the ever-increasing number of trailer lorries on the main road into town and through the town centre. The 'noises, smells and dust' that come with the new goods transportation are new and have not been met with the (almost) content attitude that iron ore transport enjoyed in its days of prosperity.

Similarly, people who are not in a leading position do not talk much about the Triangle. Other matters are more important, according to those we interviewed. Most strikingly, people are concerned about and regret the public quarrel that is going on among the local leaders. Otherwise, attention is spread over many subjects, but infrastructure preoccupations shine through as particularly 'hot issues': for many years, inhabitants have fought for the life of the secondary airport, built in the 1970s, which is located within the town. Today, most people seem cautiously and reluctantly ready to consent to its closure since, instead, an analogous fight for a bridge construction is about to succeed. The new bridge will reduce the travelling time to the main regional airport from 64 minutes to 49 minutes (Norwegian Public Roads Administration/Northern Region 2005).

In other words, the inhabitants of the town of Narvik, who used to be 'in front' with the train, have lately focused their attention towards the new supreme symbol of modern life, which is air travel. The recent infrastructural preoccupations suggest that the meaning of train-engendered mobilities has changed, and that the new mobilities of air travel are involved in the production of place. Interviews tell us that the 'imagined mobilities' (cf. Cresswell 2006) of train journeys are referred to in a retrospective manner. Regionally, the train is still unique to Narvik, but it is no longer a superior mobile quality that can underpin an experience of being 'important' or enjoying first-class access to more 'modern' lifestyle opportunities. In addition to this, the inhabitants make use of an interstate road that opened in the 1980s and enjoy the freedom and flexibility of travelling to Sweden by car. In contrast to the past, these new meanings of mobilities do not provide broad popular support for the interests of the transport actors, even though the new transport actors are gaining from all the new infrastructure constructions mentioned.

Concluding Remarks

The case of the town of Narvik and its transport industry describes a particular relationship between place and mobilities, articulated distinctly at different moments in history. This is not a dichotomy relation between the 'static' place and the 'dynamic' mobilities: rather, our account tells of how practices of mobilities provide experiences, imagination, power and physical materialities that are a fundamental part of the continuous social process where a place is produced. These descriptions also underline the importance of conceptualizing place as extrovert, and thereby open to accounts of place as '...articulated moments in networks of social relations and understandings...' (Massey 1994, 154). In addition to this, the raw material industry-related economy that has marked the town of Narvik in particular ways from the start has established evident external relations; the social reproduction of the place circulates within a space that clearly reaches beyond the physical borders of both place and nation (Friedman 2007). This chapter also tells of how mobilities relating to goods and raw material transport have challenged place in particular physical-material and aesthetic ways: the space-consuming character of transport, the huge installations required and their environmental implications may stretch the tolerance of the surroundings. As such, the transport industry is likely to exert a considerable influence on place negotiations.

The 'power geometry' within which the town of Narvik has been positioned shows that mobility and power are not only a matter of being able to move or not, but also involve a question of controlling flows (Massey 1994, 149). In the case of Narvik, LKAB held this control. Inspired by Foucault's conceptualization of power as a characteristic of social relations, as something which people *do* (and do not *have*) (Foucault 1982), one might say that the townspeople experienced a dependency on the flow of iron ore through its centre. The inhabitants were dominated by LKAB, and partly also by the main landowner and infrastructural provider, NSB. The flow-through did provide wealth and pride, but space for opposition was limited.

Furthermore, the recession entailed a 'separation' of the space-consuming transport industry and the town. The inhabitants were pushed to establish independence from the iron ore industry. The dawning optimism that is now to be found is also the fruit of such efforts. The 'separation' has produced a town-life containing the diverse qualities we have described in this chapter. These qualities correspond with many of the features of the 'post-industrial space' described by Lash and Urry (1994, 216–17). The present context provides unfortunate circumstances for the transport industry actors who are once again insisting on space. The antagonism displayed lately bears the marks of a social relational production of place that has rather unnoticeably lost its former dominant actors, and where negotiations are open to confrontation.

When, more than one hundred years ago, modernity so loudly and uncompromisingly reached Narvik, the iron ore transport actors overwhelmed this small tongue of land by the fjord. The roots of the transport industry's dominant position have been severed, however, as de-industrialization has taken place. 'Reindustrialization' of space, advancing within today's Northern European peripheries (Nyseth and Granås 2007), potentially raises the heat of place negotiations to new heights.

References

Aas, S. (2001), *Narviks historie, bind 1. Byen, banen og bolaget* (Narvik: Stiftelsen Narviks historieverk).

Bauman, Z. (2000), 'Time and Space Reunited', *Time and Society* 9: 2–3, 171–85.

Cresswell, T. (2006), *On the Move* (New York: Routledge).

Dreyfus, H.L. and Rabinow, P. (eds) (1982), *Michel Foucault: Beyond Structuralism and Hermeneutics* (Chicago: The University of Chicago Press).

Foucault, M. (1971), 'Nietzsche – genealogien, historien', *Arr: Idehistorisk Tidsskrift* 2:1, 4–13.

—— (1982), 'The Subject and Power', in Dreyfus and Rabinow (eds).

Friedman, J. (2007), 'Global Systems, Globalization and Anthropological Theory', in Rossi (ed.).

Lash, S. and Urry, J. (1994), *Economies of Signs and Space* (London: Sage).

Massey, D. (1994), *Space, Place and Gender* (Cambridge: Polity Press).

—— (2005), *For Space* (London: Sage).

—— (2007), 'Is the World getting Larger or Smaller?', *openDemocracy* [web site], (updated 15 February) www.opendemocracy.net/globalizationvision_reflections/world_small_4354.jsp.

Norwegian Public Roads Administration/Northern Region (2005), *Sammendrag av konsekvensutredning for transportkorridor E6 Narvik-Evenes* (Tromsø: Norwegian Public Roads Administration/Northern Region).

Nyseth, T. and Granås, B. (eds) (2007), *Place Reinvention in the North. Dynamics and Governance Perspectives* (Stockholm: Nordregio).

Rossi, I. (ed.) (2007), *Frontiers of Globalization Research: Theoretical and Methodological Approaches* (New York: Springer-Verlag).

Simonsen, K. (2001), 'Rum, sted, krop og køn. Dimensioner af en geografi om social praksis', in Simonsen (ed.).

—— (ed.) (2001), *Praksis, rum og mobilitet. Socialgeografiske bidrag* (Frederiksberg: Roskilde Universitetsforlag).

Svendsen, O. (2002), *Narviks historie, bind 2. Storhetstid, brytningstid, framtidshåp* (Narvik: Stiftelsen Narviks historieverk).

Transportutvikling Ltd. (2004), *The Northern East West (N.E.W.) Freight Corridor. A Global Trading Route* (Narvik: Transportutvikling Ltd.).

Chapter 16

Politics of the Interface:
Displacing Trans-border Relations

Ari Aukusti Lehtinen

Introduction

In December 2006, the Russian Ministry of Natural Resources founded a large National Park in the Karelian Republic, close to the border with Finland, north of the town of Kostomuksha. The establishment of the Kalevala National Park, consisting of 74,400 hectares of mainly old-growth forests, is a remarkable environmental signal and advance, as it effectively marks the end of a seven-year period of low profile in Russian nature conservation terms. This decision was the result of persistent collaborative work, both official and non-governmental, across the border and it has been long and anxiously awaited by Russian and Finnish forest conservationists. The Kalevala Park had remained high on their priority lists for 15 years, since the launch of the Greenbelt Programme, aimed at protecting the remaining old-growth forests on the border.

The Kalevala Park has also been welcomed by the local people, as it is perceived as a guarantee of new jobs for the area. The running of the Park's administrative centre will bring dozens of new jobs to Vuokkiniemi, the central village of 500 inhabitants near the Park, and the village will benefit in general terms from new social and entrepreneurial activities linked, for example, to increasing trans-border tourism. The cultural associations on the Finnish side of the border have also widely supported the Park, as it protects many of the forests surrounding the settlements where significant parts of the *Kalevala*, the Finnish national epic, were collected by Elias Lönnrot during the 1830s and 1840s. *Vihreä lanka*, the Finnish weekly paper of the Green Party, for example, headlined the Park news in a telling way: 'Russia conserves the roots of the Finns' (Laitinen 2007).

The Park decision is an example of successful trans-border co-networking of a diverse compilation of hobby and lobby groups. Together, they were able to exert a remarkable influence on the socio-spatial re-organization of the border zone during a period of geopolitical transition. Largely due to their efforts, the closed border of the Cold War has turned into passages of cultural and ecological co-operation. The most marginal backwoods were revalued as hot spots of biodiversity conservation. The border villages, long under pressure to be emptied, gained value as lively centres of heritage conservation. The logging companies gave way to wilderness developers. How and why did the conservationists succeed in mobilizing and legitimizing the idea of the border as a Greenbelt? What were the transforming and transformed processes like? This chapter aims to provide answers to these questions.

Relational and Concentric Spaces

This chapter focuses on the above-mentioned transformation of Finnish-Russian border circumstances by relying on processual ideas of place-making and community building, and using this ground to develop an approach that is sensitive to both general dependencies and local particularities of the Belt. Communities of the Belt, both local and international, are seen to have evolved through three interrelated spheres of co-development. Firstly, societal change is understood to have resulted from locally specific responses to the pressures of globalization. Places, localities and communities function, according to this view, as intersections or moments of wider trajectories of multinational networks (Massey 2005; see also Chapter 13 by Jóhannesson and Bærenholdt). Secondly, socio-cultural change is notified in transformations of communicative conditions, especially in the choice of language and acts of renaming. Socio-cultural identification and estrangement are integrally linked to lingual practices and this makes them delicate political issues in trans-border circumstances (Tanner 1929; Lehtinen 2006). Thirdly, changes in imaginary orientation are identified in the transformation of the central socio-spatial myths and metaphors upon which the communities more or less purposefully depend in their daily routines (Connerton 1989; Buttimer 1993; Buttimer and Mels 2006).

This type of layered interpretation of socio-spatial change underlines the dependent and tensional relations between the three spheres of the Belt. These tensions often emerge in contests between the general relational pressures of continuous displacement and particular acts of community relocation, according to inherited and adopted patterns. Human communities, stretching from neighbourhoods to globally networked allies, are accordingly seen as evolving along two complementary paths of co-development. Firstly, they progress reactively, through cumulative effects emerging from within general changes in society (Vähämäki 2000; Virno 2004). This change may be termed relational displacement. Secondly, human communities evolve proactively, through a transformation of shared memories and collective practices grounded in the experience of belongingness: that is, through concentric initiatives and imaginary relocation (Jürgenson 2004; Knuuttila 2005; 2006; Schwartz 2006). Relational and concentric realms of change thus serve here as complementary socio-spatial categories, or development paths, both of which need to be recognized when aiming to comprehend the meeting of the general and the particular in the Belt communities.

In this perspective, communities appear as socio-spatial processes, where the domains of the familiar are constantly challenged by temporally and spatially diverging flows of competing interests and mixtures in their meetings (see Kymäläinen 2006). Local communities, for example, cannot avoid the changes caused by their reactions against those intervening factors which are categorized as menacing. These defensive reactions, aiming to confirm the dividing line between the secure interiority and the threatening exteriority, keep the communities in movement. This is a socio-spatial process that gains its impetus from the unruly consequences of community relocation. Communities, as well as their defenders, can become radically renewed when being attacked by marginal but counted-as-threatening signals from beyond the familiar order (Lehtinen 2006, 215–34).

However, this way of reacting to intervening factors is also strongly connected to the inherited socio-cultural and imaginary dynamics of community building. Community change includes transformations of socially performed and habitual messages, or memories, which are slow in their renewal but which can also be rapidly reorganized in thorough socio-cultural or imaginary displacements (cf. the discussion of taskscapes in Chapter 3 by Mazzullo and Ingold). Paul Connerton (1989) examines in detail ritually performed images of the past that serve, in the form of implicit narratives, as supporters of the dominant social order. He thus argues that this way of sharing memories produces communities where remembering takes place. Profound changes in society occur, as Connerton (1989, 6–13) argues, through the replacement of central performative practices and their readings, celebrating the implicit narratives of the community (see also Kymäläinen and Lehtinen 2007).

The relational emphasis, when considering communities as intersections of general projections, carries the risk of overlooking the multiple human co-associations in space and time. Ignoring the concentric side of human community building can easily result in a lack of sensitivity to emerging new initiatives from the margins of society which deviate from the generally accepted framework. Multinational interdependence, when viewed as an all-encompassing denominator of socio-spatial variation, understates the innate dynamics of communicative and imaginary interaction. The circles of particularity are then simply ignored, or judged to be uninteresting, since the questions binding them have no general bearing.

Communities may thus be viewed as particular actors of multilayered change. The contested co-formation of concentric and relational spaces, combined with a proactive stance in building personal and collective coping strategies, is now the founding forum and event of societal critique. This is the place of socio-spatial empowerment. Community relocation becomes associated with an appreciation of everyday conditions and settings where people and their beliefs meet, and it also covers the processes that these meetings co-produce (see Chapter 17 by Benediktsson). Individual communities are thus not regarded as particular products of society, developed through their reactions to signals from the 'outside', but are instead accompanied and co-informed by chains of memories and customs that respect a shared past. The Greenbelt, for example, is a zone of forest 'hot spots' under negotiation between forest industry and conservationists, but it is also a cultural contact field, or an imaginary interface, for several diverging cosmologies and chains of border memories.

The Greenbelt is analyzed below with the help of the above-formulated community approach, by emphasizing the process-type dialogue between socio-spatial displacement and relocation, and thus paying attention to both concentric and relational dimensions of community-making. The analysis starts with a description of the societal formation of the Belt proposal. This is followed by an excursion into changes in the specific socio-cultural conditions of the communities affected by the Belt proposal. The final part of the chapter describes how socio-spatial myths and metaphors function in border co-operation.

Societal Change: From a Dividing Line to a Local-global Programme

The Greenbelt is a political programme, or proposal, that was launched in the early 1990s by nature conservationists and by the Ministries of the Environment in Finland and Russian Karelia. The Belt co-operation was initially motivated by worries concerning the fate of the border forests in the Republic of Karelia (see Pyykkö 1994; Tikkanen 1997), and since then Russian Karelia has been the main target, and challenge, for the Belt proponents.

The border forests of Murmansk and Leningrad oblasts and Eastern Finland soon became integrated in the process and the Greenbelt of Karelia consequently grew into the Finnish-Russian Greenbelt. It was moreover retitled at the beginning of the twenty-first century by ecologists as the Fennoscandian Greenbelt, a definition that underlines the existence of a particular Eastern Fennoscandian vegetation type, *Florae Fennicae*, which is dominant on both sides of the border. The Fennoscandian emphasis has also made it easier to list the eastern border areas of Northern Norway under the 'umbrella' of the Belt. The territorial-ecological concept of Eastern Fennoscandia covers most of Finland and the Kola Peninsula, and it extends to the Onega and Ladoga Lakes in the south-east. It is carefully distinguished as a separate region from the more continental vegetation areas of inland Russia. According to Raimo Heikkilä and Tapio Lindholm (2003, 11), two Finnish ecologists active in Greenbelt promotion, the Belt 'belongs to the Fennoscandian Archaean bedrock area with gently undulating terrain formed of several glaciations during the latest two million years'.

The Fennoscandian Greenbelt has also been connected to the conservation of the border Belts running through the Baltic countries and Central Europe over to the shores of the Mediterranean (Fanck et al. 2003, Schwartz 2006). This chapter, however, concentrates on the Belt co-operation between Finland and Russia. The abbreviation Belt thus refers here to the Finnish-Russian Greenbelt (Figure 16.1).

According to nature conservationists, the Belt connects the continental Russian taiga with the more oceanic boreal forests of Finland and Scandinavia. The fragmented islands of protected forests on the western side of the border are fuelled by species migration from the 'genetic mainland' in the east. The Finnish-Russian Belt, extending over a thousand kilometres from the Gulf of Finland up to the Barents Sea, also bridges several vegetation zones within its north-south axis. This is perceived as being an important ecological factor, forming a natural corridor for the species to adjust to ongoing global climatic changes (Pyykkö 1994).

The Belt is generally valued as a zone of conifer-dominated forest that has never been exploited by large-scale forestry. This is due to the fact that throughout historical times, the border forests have remained in the margins of settlement pressure in Northern Europe. Military purposes during the Cold War also secured the zone from logging operations and moreover, some of the westernmost villages in the Soviet Union were emptied or, in Russian terminology, liquidated. These evacuations lessened the already minimal pressure on the surrounding nature. The natural(ized) history of the Russian side of the Belt has left species and ecosystems on their own, only to be later found and identified as remnants that had largely disappeared from the well-managed forests of Finland and Scandinavia.

Figure 16.1   The European Greenbelt
Source: by Timo J. Hokkanen.

The Finnish-Russian Belt is an anomaly in between two different territorial jurisdictions, and it is also an interface of several ethnic identities and cultures. The border regions and communities share a common history of dramatic geopolitical regime changes. The state border thus functions as a societal fact that divides the communities, even though in socio-cultural terms it promotes a sense of neighbourhood, reflected for example in the intense lay co-operation during periods of easy access across the border.

The Greenbelt also symbolizes the fact that, despite their deep historical societal contrasts, the East and West of the European North have grown closer and closer to each other. Northern Europe, extending from the shores of the Atlantic to the Ural Mountains, has in many respects become a single arena for environmental and forest co-operation. This is particularly true in forest conservation terms. Greenpeace, Taiga Rescue Network and the World Wildlife Fund have all extended their activities across the forests of Northern Europe. For them, the question of old-growth forests connects and unifies the eastern and western sides of the European North. The Greenbelt is a symbolic and practical (geo)political element in this unification (see Lloyd 1999).

The gradual political development of the Belt also reflects broader changes in global north-south co-operation in forest utilization. The options of fast-growing eucalyptus and acacia trees in tropical and temperate plantations have already challenged northern forestry practices, which for a long time have favoured pulp and paper production. This new competition setting has forced an active search for alternatives to producing northwoods, primarily for paper.

The conservationists of the Greenbelt have strategically linked the new global signals of forest trade with the civic pressure of environmentalism in Europe. Old-growth forests are valued highly in the urban centres of the European Union and there is no reason to believe that this concern is limited to nature conservation. The old-growth forests of the North are tourist attractions and future targets for urban Europeans searching for second homes in primeval nature. In addition, the various forms of traditional knowledge on the part of the Belt communities have been valued by Belt protagonists as a sound strategic alternative to the polarized contest between industrial exploitation and conservation. This is especially the case in the Kalevala villages and among the Sámi communities at the northern end of the Belt (Nykänen 2006; Linjakumpu and Valkonen 2006).

The emergence of the global South as a factor in the Greenbelt development has thus forced the Belt actors to rethink their societal dependencies. It has become difficult to define the Belt as a narrow zone on the map any longer. Instead, it seems reasonable to identify the Greenbelt as a network of actors and factors linked by shared or colliding interests. The Belt extends from Upper Lapland to urban Europe and stretches its 'fingers' to Southeast Asia and also to the Amazon Basin. The Belt is a matter of negotiations among industrial developers and consumers, local and union activists, indigenous networks, landless groups and nature conservationists in different parts of the world. Changes in one part of the network have an impact on other parts. The Greenbelt is made up of societal networks sharing and dividing limited forest resources, both local and global.

Socio-cultural Change: From Marginal Reactivity to Multilingual Pro-activity

The zonal-like Greenbelt was initially introduced as a macro-scale societal programme, primarily comprehensible in the maps and satellite views of Northern Europe. The launch 'from above' was reminiscent of the geopolitically turbulent past at the border and therefore raised tensions in the local communities. Depending on the perspective chosen, the Belt was seen as a limit to local activities or as a new source of livelihood and identity (Hokkanen 2000; Nykänen 2006). The setting underlines the critical local-global relations between those benefiting from the Belt decisions and those suffering from them. The decisions tend to rank individuals and communities according to their location and flexibility in the new communicative networks.

The border conditions of the Belt form a particular case of identity politics, as several territorial lingual regimes overlap the local lifeworlds. Finnish, Karelian, Russian, Sámi and Scandinavian languages and dialects meet at the border communities and the main cultural divide runs between the Finno-Ugric and Indo-European languages. The former are spoken by small and practically stateless ethnic minorities, such as the Sámi, the Karelians, the Komi, the Mansi, the Hanti and the

Samoyeds, but also by some groups more or less secured by a state administration of their own, such as the Estonians and the Finns. The main dividing line becomes manifest not only in daily difficulties in coping with strange words but also in coping with strange worlds. In addition to hardships in practical communication, the overcoming of lingual borders always involves a reflection upon the distance between two radically differing systems of signification: Finno-Ugric and Indo-European. This means that the worlds not only look different across the lingual border: they are in fact different.

The most local and indigenous ethnic communities of the Belt have, for geopolitical and historical reasons, found it necessary to learn several neighbouring languages; it is, for example, not rare to meet true polyglots in the northern end of the Greenbelt who are able to communicate through a whole spectrum of local-global languages, ranging from several Sámi languages (out of the total of nine) to their Scandinavian and Finnish counterparts, and sometimes also extending to Russian. Due to the needs of tourism entrepreneurship, English is also a necessity for many today. The polyglot communities from the extreme margins look at the surrounding worlds from a multilingual perspective, comparing the surrounding power regimes and cultural orders from various complementary angles.

The multilingual configuration of the Belt is, however, not something fixed or unchanging. The polyglot skills are continuously challenged by the same societal pressures from which they were initially forged. The historical dynamics of the Belt are reflected in the variation of local and generational dialects, as well as in the multi-ethnic mixtures. Societal changes have also resulted in violent transformations and compromises in the social life of the border communities, in the form of forced and voluntary evacuation and resettlement. Regime changes have consequently had a deep impact on the practices of domination, both in terms of governance and in the formulation of expertise. The tension between authorized and lay forms of communication has also remained at the practical core of Belt politics. Expert language dominates the nature conservation agenda and only occasionally, as in the case of Vuokkiniemi and the Kalevala National Park, has local confusion turned into fluent co-operation. The local success of the Kalevala Park shows that contact is possible, with the help of translators who have devoted considerable time to negotiations across the division of professional and lay regimes (Nykänen 2006).

Imaginary Change: Metaphorical Reorientations

The success of the Belt programme is related to how thoroughly the 'Belt ideology' reaches the minds and thoughts of the actors participating in the Belt development. The degree of awareness and willingness on the part of the actors is critical to the discursive strengthening of the Belt projection. Socio-spatial metaphors play a central role in the promotion and transformation of Belt initiatives. The Belt is, for example, a key metaphor for nature conservationists and helps us all to shape the border forests as an ecological corridor. This is, however, ecologically and politically problematic, bearing in mind the fragmented character of the border forests that are realistically worth conserving.

The Belt initiative has reanimated several of the historical signposts of cultural orienteering at the border, mobilized for ideological and political purposes in cross-border contacts. Something about the imaginary dynamics of the Belt, and border, is recognizable from the specific socio-spatial expressions of re-membering and othering in community-building, ranging from connotative expressions about neighbourhood relations to a routine-like practice of necessary skills which help individuals and communities to cope with the northwoods. Some imaginary geographies linked to contests between central border metaphors, such as the bear, Siberia, the North and wilderness, will be discussed below.

Greenbelt protection in the European Union is explicitly connected to the protection of the remains of the boreal megafauna, such as wolves and bears. The bear has also been a central liminal figure over the centuries, clothed with a variety of euphemisms in Finno-Ugric identifications. It is loaded with a mixture of respect and fear. In many historical imaginations the bear has been valued as a close relative to the northern peoples, and this attitude can still be identified in some rituals and taboos connected, for example, with bear hunting (Klemettinen 2002). By contrast, the almost human guardian of the northern forests is, for many, also a merciless beast of the wilderness. In Eastern Finland, for example, a deep-seated fear of the bear, or *petoviha* (wrath for the beasts), may be perceived as a partial explanation for locally popular attitudes against forest conservation (Palviainen 2000). It is feared that the Belt might become the territory of the beast, a home area for a terrorist from nature, learning to live close to human settlement and regularly visiting border villages. The Belt also reminds the Finns of their close relationship with their eastern neighbour, the Russian Bear.

The low administrative status of Karelia in Russia, on the other hand, demonstrated for example in the daily hardships of the emptying forest villages (Varis 1996), exemplifies general Federal attitudes towards the development potential of the Karelian backwoods. The border settlements of the Karelian Republic belong in the imaginative geographies of the Federal administration to a metaphorical Siberia that symbolizes the long distance from the capital city. Siberia nearest to Moscow is part of the antipodes to success of the careers of civil servants (Klementjev 1990). The Belt villages are perceived as having no serious role to play in the development of the Russian economy as a whole. Those exiled from Moscow to the Belt margins, as if by penal measure, lack motivation or the potential to initiate any local or trans-border projects by themselves. The metaphorical load of Siberia is also familiar to the communities on the Finnish side of the border, grounded in historical narrations of criminals sent to distant eastern parts of the Russian Empire to which Finland belonged in 1809–1917. These narrations are alive and well in the modifications of a Finnish saying about the Siberian way of teaching: '*Kyllä Siperia opettaa*' ['Siberia will teach you'] mirrors the sceptical attitudes toward newcomers with strange ideas.

Moreover, the emphasis of the northern particularity of the Greenbelt, explicated for example by Belt descriptions that are based on the impact of glaciation and the proximity of permafrost, has also largely supported the value horizons of the local border communities. The North, as a metaphorical dimension, surfaces in local notions of distinctness and it also functions as a source of pride. It is essential to learn the rules of the North in order to survive in the backwoods. The skills of the North are a means of self-identification, and they also function as tools to distance oneself from those newcomers from the South. The practical skills of coping with the North serve as a way of measuring one's 'Nordicity'. The imaginary North can thus be used occasionally as a powerful tool of discrimination.

Finally, Greenbelt conservationists were also able to mobilize the idea of wilderness in the Greenbelt negotiations. Some border locations have been turned into hot spots of wilderness tourism, whilst others have become ethical issues for critical paper consumers in the European Union. The fate of *Urwälder* and the wildernesses has grown into key issues of Belt politics and the development has challenged the local practical evaluation of the same areas as everyday landscapes. Wilderness strategies, when initiated as map projections 'from above', and when resulting in strict Park decisions, may appear as serious limitations to local modes of living (see Heikkilä 2006).

Conclusions

This chapter has distinguished three layers, or spheres, of socio-spatial change relating to Greenbelt politics, namely societal, socio-cultural and imaginary (metaphorical). The societal change, that is the politico-economic formation of the Belt network, has been perceived as being grounded in ambitious ecological conservation interests, which were modified in turn by a wide spectrum of responses to them. On the other hand, socio-cultural transformations, including multilingual re-combinations, politics of naming and tensions between expert and lay vocabularies, have been perceived as central means of identity-building. These issues surface, for example, in negotiations between developers and traditionalists, but also in dialogues across generations. Finally, the imaginary sphere of metaphorical relocations has emerged as the realm that re-interpreted the symptoms of rapid societal change within the framework of shared memories and routines. This sphere was also easily mobilized for the purposes of socio-cultural discrimination, especially during periods of unruly societal renewal and displacement.

The two last-mentioned layers of socio-spatial change were partially located in the sphere of internally shared and negotiated communication, both explicit and non-reflexive, which constituted what might be termed the concentric features of socio-spatial development. Accordingly, places, localities or communities were not only perceived as intersections or moments of wider multinational trajectories: they were also valued as having a proactive dynamic of their own. The Belt excursion has offered us views on border communities which, over the centuries, have been thoroughly modified by multinational forces. However, as became apparent during this process, the violent geopolitical past has not fully wiped away the particular expression of local histories.

The distance between the local premises of the border people and multinational Belt thinking was gradually overcome in the long process of founding the Kalevala National Park. Developing contacts demanded considerable efforts on the part of the 'translators', due to the initial ideological framework of the Belt proposal. The

Greenbelt was, and is, a primarily political construction that aims to protect the last remaining old-growth forests at the border zone. The Belt is an ideal construction, a purified representation of nature, launched to make us recognize the Belt shape of the individual forest areas close to the state border which are worth conserving. The tendency to construct purified concepts in nature conservation is indelibly linked to the long Western tradition of park conservation that has its roots in the idea of indigenous nature needing to be protected from human modification. The original exclusion of humans is the threshold that burdens the 'translators' wherever the parks and belts overlap with human communities.

The purified Greenbelt has also appeared as a means of contestation in place-marketing. The unpolluted and monumental nature of the North has been used as a tourist attraction for a long time (Birkeland 2005; Pitkänen 2005), and an extended history of marketing campaigns has resulted in a multitude of co-constructions of imaginary scenarios that have been widely promoted by the locals as well. The purified presentations have in a way successfully obscured the contamination of the daily environment of the North. The Greenbelt may thus be viewed as a mask, for example, a monument of almost pristine nature that hides actual transformations of border conditions, such as intensive forestry, peat draining and environmental pollution.

The specific history of Belt naming is also an expression of the power of purified representations of nature. The re-emergence of the vegetation and bedrock maps of Eastern Fennoscandia in Belt politics are loaded with geopolitical connotations that have their roots in the scientific mapping of the natural extension of Finland in the East during the pre-independence decades and the Second World War period (see Paasi 1990). The Finnish-Russian launch of the Belt was, however, co-operative and should have been taken as a clear signal to limit any forms of Fennoscandianism in the Belt politics. The setbacks and slow progression of the Belt conservation during the early years of the twenty-first century in Russia may have something to do with the radical re-naming of the Belt, but the question relates fundamentally to our sense of the past: learning from the past can only be grounded in a clear sense of the historical socio-spatial transformations, including the diverse imaginations and ambitions attached to them. This is also the methodological message of this chapter: the learning associated with the concentric side of Belt change can only be grasped by sharing the routines and memories of the border people – in close contact with the drama and the dynamics of human communities coping with these daily questions.

The methodological lesson thus suggests that the concentric elements of lingual and habitual practices can no longer be ignored or treated as 'black boxes' of contingency, when aiming for a full understanding of the layered dynamics of place-making in the era of neo-liberal time-space compressions (cf. Chapter 4 by Birkeland). Ignoring them is easy, due to the form of concentric communication which consists of signs and practices that are non-existent or uninteresting to those who have not learned them through daily routines (see Ingold and Kurttila 2000). This chapter has also shown that sensitivity to the more unarticulated spheres of community-building may significantly help us to deepen our views on the historical divergence of social space and power. The politically generative side of community-making is firmly related to the potential of individual actors anchored in their multilayered socio-cultural relations and imaginary projections; those of us studying socio-spatial politics or the politics of place clearly cannot afford to overlook this source of community empowerment.

References

Birkeland, I. (2005), *Making Place, Making Self* (Aldershot: Ashgate).

Buttimer, A. (1993), *Geography and Human Spirit* (Baltimore: Johns Hopkins University Press).

Buttimer, A. and Mels, T. (eds) (2006), *By Northern Lights* (Aldershot: Ashgate).

Connerton, P. (1989), *How Societies Remember* (Cambridge: Cambridge University Press).

Fanck, M., Frobel, K., Geidezis, L. and Graf, L. (2003), *Germany's Green Belt* (Bund: Nurenberg).

Häyrynen, S. (ed.), *Kulttuurin arviointi ja vaikutusten väylät* (Helsinki: Cupore Publications 12) 51–65.

Heikkilä, L. (2006), *Reindeer Talk* (Rovaniemi: Acta Universitatis Lapponiensis 110).

Heikkilä, R. and Lindholm, T. (2003), 'The Nature Reserve Friendship as part of the Fennoscandian Green Belt', in Heikkilä and Lindholm (eds).

—— (eds) (2003), *Biodiversity and Conservation of Boreal Nature* (Kajaani: Kainuu Regional Environment Centre) 11–12.

Hokkanen, T. (2000), *Ympäristöjärjestöt ja kansainvälinen metsäteollisuus kiistakumppaneina Karjalan tasavallassa*. Master's thesis, mimeo (Joensuu: University of Joensuu, Department of Geography).

—— (ed.) (2006), *Can the Interests of Forestry, Local People and Nature Conservation be Combined?*, 21 (Joensuu: North Karelia Regional Environment Centre).

Ilomäki, H. and Lauhakangas, O (eds), *Eläin ihmisen mielenmaisemassa* (Helsinki: Suomalaisen Kirjallisuuden Seura) 134–73.

Ingold, T. and Kurttila, T. (2000), 'Perceiving the Environment in Finnish Lapland', *Body and Society* 6: 3–4, 183–96.

Jürgenson, A. (2004), 'On the Formation of the Estonians' Concepts of Homeland and Home Place', *Pro Ethnologia* 18, 97–114.

Klementjev, J. (1990), 'Autioituvien kylien ikävä tarina', *Punalippu* 64:2, 88–96.

Klemettinen, P. (2002), 'Kurkistuksia karhun kulttuurihistoriaan', in Ilomäki and Lauhakangas (eds).

Knuuttila, S. (2005), 'Samankeskinen paikallisuus – rajallisen kiinnostuksen autenttisuus', in Häyrynen (ed.).

—— (2006), 'Paikan moneus', in Knuuttila et al. (eds).

Knuuttila, S., Laaksonen, P. and Piela, U. (eds) (2006), *Paikka. Eletty, kuviteltu, kerrottu*, 7–11 (Helsinki: Suomalaisen Kirjallisuuden Seura).

Kymäläinen, P. (2006), *Geographies in Writing. Re-imagining Place* (Oulu: Nordia Geographical Publications) 34: 3.

Kymäläinen, P. and Lehtinen, A. (2007), 'Places on the Move. Rethinking the Chora of Geography', a manuscript (Mimeo).

Laitinen, J. (2007), 'Venäjä suojelee suomalaisten juuret', *Vihreä Lanka* 4, 6.

Lehtinen, A. (ed.) (2005), *Maantiede, tila, luontopolitiikka. Johdatus yhteiskunnalliseen ympäristötutkimukseen* (Joensuu: Joensuu University Press) 135–60.

—— (2006), *Postcolonialism, Multitude, and the Politics of Nature* (Lanham: University Press of America).

Linjakumpu, A. and Valkonen, J. (2006), 'Greenpeace Inarin Paadarskaidilla – verkostopolitiikkaa lappilaisittain', *Politiikka* 48:1, 3–16.

Lloyd, S. (ed.) (1999), *The Last of the Last: The Old-growth Forests of Boreal Europe* (Jokkmokk: Taiga Rescue Network).

Massey, D. (2005), *For Space* (London: Sage).

Nykänen, R. (2006), 'Local People and Tourism: Working Towards Genuine Locality and Good Quality', in Hokkanen (ed.).

Paasi, A. (1990), 'The Rise and the Fall of Finnish Geopolitics', *Political Geography Quarterly* 9:1, 53–65.

Palviainen, S. (2000), *Suurpedot Pohjois-Karjalassa* (Joensuu: Pohjois-Karjalan liitto).

Pitkänen, K. (2005), 'Kesä Suomessa. Semioottinen tulkinta 2000-luvun järvimatkailumaisemasta', in Lehtinen (ed.).

Pyykkö, J. (1994), 'The Green Belt of Karelia at Risk', *Taiga News*, 10, 6–7.

Schwartz, K. (2006), *Nature and National Identity after Communism* (Pittsburgh: University of Pittsburgh Press).

Tanner, V. (1929), 'Antropogeografiska studier inom Petsamo-området. I. Skoltlapparna', *Fennia* 49, 1–518.

Tikkanen, O-P. (1997), 'Karjalan vihreä vyöhyke', *Alue ja Ympäristö* 26:1, 60–67.

Varis, E. (1996), *The Restructuring of Peripheral Villages in Northwestern Russia* (Helsinki: Wider Insititute).

Virno, P. (2004), *A Grammar of the Multitude* (New York: Semiotext(e)).

Vähämäki, J. (2000), 'Displacing the Means of Political Control', Hänninen, S. and Vähämäki, J. (eds) *Displacement of Politics* (Jyväskylä: SoPhi) 46–64.

Chapter 17

Moving Places:
The Emotional Politics of Nature

Karl Benediktsson

Introduction

All of a sudden, the politics of nature have moved centre stage in Iceland. On the whole a rather placid and contended nation, usually more preoccupied with petrol prices than with philosophical introspection, Icelanders have found themselves embroiled in deeply divisive battles over the meaning of their own nature and how it could and should be used to fashion desirable futures for the country. These battles have centred on the building of large-scale power stations and the concomitant industrial strategies pursued relentlessly by the state and other actors. In the process, some widely differing cultural values and constructions of nature have been exposed. At the centre is an entanglement of various mobilities – of people, images and ideological convictions, not to mention capital – and how they constantly shift and rework the content and meaning of place.

This is, in a sense, a *déjà vu* in the Nordic context. Few if any clashes in Nordic cultural politics have been as emotionally charged as those relating to hydropower projects. In around 1980, the Alta conflict in Northern Norway brought issues of indigenous rights sharply into focus. Precisely such issues may be absent in Iceland, but in their stead has come an increasing realization on the part of the Icelanders themselves that, in the early twenty-first century, theirs is but a small place, which is being subjected to various international forces and the character of which is being altered more rapidly than at any time before in its history. 'Nature' has become fundamental to the Icelanders' sense of themselves (cf. Hálfdanarson 1999) – and their sense of the place in which they happen to live.

This chapter has its geographical focus in East Iceland, the current site of an extraordinary re-engineering of nature and place – the Kárahnjúkar project. It first traces the international changes in the conservation discourse, which have mobilized the nature of certain places, so to speak, and made East Iceland a battleground where globally significant 'tournaments of value' (Appadurai 1986) take place, involving ethics and aesthetics, as well as economic and other issues.

I take my theoretical cue from recent arguments about nature's inevitable sociality – and *vice versa*. Macnaghten and Urry (1998) argue against three common 'doctrines' of nature – what they term (environmental) realism, idealism and instrumentalism. Instead of these doctrines, they emphasize the need to engage with complex, 'embedded' social practices that 'produce, reproduce and transform different natures and different values' (Macnaghten and Urry 1998, 2). This is a well-rehearsed theme in recent scholarship about nature-society relations (cf. for example, Braun and Castree 1998; Castree and Braun 2001; Whatmore 2002). 'Natural places' are anything but 'natural': they are relationally produced and reproduced locally, through embedded practices resulting from the entanglement of these places in economic and social processes, some of which are global in their reach. As such, the stories I shall tell are thus illustrative of 'the production of socio-natures on a "global" scale, but in ways that are remarkably differentiated, a globalization of nature that is spatialized in particular and consequential ways' (Braun 2006, 647). We shall look – very selectively – at some aspects of this process in the Kárahnjúkar project in East Iceland. Mobility plays a central role in all of them, albeit in varying ways.

Articulating Nature for Industrial Development

At the outset, however, it is necessary to trace how the nature of the highland plateau of Iceland came to be articulated initially as places for accumulation, or configured for industrial development, culminating in the Kárahnjúkar project. The history is long and complicated, but a short account will have to suffice.

For millennia, Iceland's numerous glacier-fed rivers – *jökulsár* – have flowed from their origins to the sea, releasing colossal energy in the process. At the beginning of the twentieth century, their harnessing for industrial use was first suggested as a viable development strategy for the then-impoverished country. The illustrious poet Einar Benediktsson, for instance, attempted to use his good connections with European élites to enrol wealthy foreigners and industrial capitalists in such plans (Friðriksson 1997–2000), but his grandiose visions were not realized until much later. Some proposals, such as the harnessing of Gullfoss – the 'Golden Waterfall' that is nowadays an obligatory stop for every international tourist in the country – were quietly set aside. Others lay dormant for some decades, only to be brought to life later, doubled and redoubled.

A watershed was crossed during the 1960s, when Iceland had its first encounter with multinational aluminium companies. After a lengthy debate and much controversy, the centre-right government of that time gingerly gave permission for the Swiss company Alusuisse to build a smelter in Straumsvík, just south of the capital Reykjavík (Skúlason and Hayter 1998). The smelter was initially rather small by the standards of the industry, but it has since been expanded several times.[1] Above all, it heralded a new era of foreign direct investment in what had until then been a very circumscribed national economy, with strong emphasis on the small scale (Jónsson 1997). It also signified a new chapter in resource development. The project was tied to the construction of a large power station at Búrfell – the first in what later became a line of power stations along Iceland's largest river, Þjórsá, and its tributary, Tungnaá, at the edge of the central highland plateau.

---

1   This smelter is now owned by the Canadian-based aluminium giant, Alcan. At the time of writing, in mid-2007, a planning proposal for yet another substantial enlargement of the smelter was recently rejected in a vote by the people of Hafnarfjörður, the town where the smelter is located.

This, of course, happened well before the advent of the truly globalized economic order in evidence today. Opposition to the first aluminium smelter came largely from representatives of the far left, who saw this as a dangerous intrusion on the part of large-scale international capital which would lead to a loss of national control over the country's resources (Skúlason and Hayter 1998). The fight against the smelter was thus fought mainly on nationalistic grounds. Nature had not yet become a serious issue of contestation.

The next inroads of heavy industry were made in the 1970s, in the form of a ferrosilicon processing plant built in Hvalfjörður, north of Reykjavík, as a joint venture by the Icelandic state and the Norwegian industrial company Elkem. There followed a period of some two decades which saw no new 'greenfield' investment in heavy industry. This was not due to any lack of interest on the part of the Icelandic government. During the early 1980s, the Ministry for Industry completed a survey of possible locations for further large-scale industrial installations, which brought attention to new sites (Staðarvalsnefnd um iðnrekstur 1983). In around 1990, a bout of intense 'smokestack chasing' nearly led to the building of a new aluminium smelter at Keilisnes, not far from Straumsvík, but the multinational consortium involved backed off (Skúlason and Hayter 1998). Finally, in the late 1990s, a new smelter was built at the same site as the ferrosilicon plant. Those industrial plants, built prior to the advent of the twenty-first century, obtained their energy mostly from Þjórsá and Tungnaá in the southern part of the highland plateau. From the power stations, the electricity pylons – growing larger with every new transmission line – marched sternly in several files across lava fields and rural districts to the industrial plants.

Reyðarfjörður was one of the places deemed suitable for the location of heavy industry quite early on. Unlike the fjords on the north coast, this tranquil fjord remains ice-free in all but the most horrible of winters (Staðarvalsnefnd um iðnrekstur 1983) – and such winters are few and far between in this day and age. The fjord itself is deep and sheltered, enabling the construction of a large industrial port at not too great a cost. Most importantly, ample power resources are to be found not far away. As early as the 1970s, the people of Reyðarfjörður – a village of some 700 people, making a living from fishing, fish processing and services – had been promised that they would get an industrial establishment in order to secure their economic base (Ólafsson 2006). The plans were relatively modest – a silicon metal factory staffed by a workforce of perhaps 130 was planned – but this did not materialize.

The large glacial rivers that flow northwards from Vatnajökull make a strong impression on anyone travelling through Northeast Iceland. During the 1940s and 50s the first ideas about harnessing their power surfaced, culminating in around 1970 with a proposal for a true 'mega project'. The idea involved a massive engineering undertaking, notably the diversion of two of the rivers towards the course of the easternmost one (Landsvirkjun 2002). Considerable research was undertaken. The scale of these initial speculative plans was such that it earned them the pithy epithet *lang-stærsti draumurinn* ('the largest dream by far') – or simply *LSD*.

Hallucinations or not, these plans resurfaced in the late 1990s in a somewhat modified version. Again, mobile industrial capital was the lure. This time it was the industrial giant Norsk Hydro, which declared itself willing to build an aluminium smelter at Reyðarfjörður. This would involve a new reservoir on the highland plateau, which would submerge Eyjabakkar – a wetland area of substantial biological importance. This soon provoked a forceful campaign among environmentalist organizations and individuals. One of the strongest points in that campaign was perhaps not so much about 'nature' as such, but about the rather obvious attempt by the National Power Company to bypass legal requirements for a fully-fledged environmental impact assessment (EIA) – in fact, permission had been issued for this power project to proceed well before the first EIA legislation was enacted in Iceland in 1994.

Due to the strong environmental opposition, the issue took yet another turn. The Ministry of Industry announced a new plan in 2000 that would leave the wetland area alone, yet allow for a much larger power station to be built by combining two large rivers into one. The juggernaut of the Kárahnjúkar project started moving. This actually amounted to several projects rolled into one, each warranting the prefix 'mega': firstly, the building of a very large dam – nearly 200 metres high – at Kárahnjúkar, together with some minor dams and other related structures nearby; secondly, the boring of some 72 kilometres of tunnels through the basalt mountain plateau (Landsvirkjun 2007) and the construction of the power station itself deep within the mountain; last but not least, the building of the aluminium smelter, which was the intended user of all this electrical energy. In addition, a new deep-water harbour had to be built and power transmission lines erected.

It is beyond the scope of this chapter to detail the intrigues of the political process and diverse actors that got the project off the ground. At any rate, it took the Icelandic environmental movement – still reeling from the Eyjabakkar conflict – some time to form a coherent reaction to these new plans. Gradually, however, the scale of the project dawned upon its supporters and opponents alike. Things really started moving: not only earth-moving machinery, with its characteristic yellow colour, but also human bodies and minds of various colours and creeds, not to mention moving representations of the highland nature that was about to be altered forever. Three stories will be told here.

First Story: Itinerant Labourers

Fieldwork, February 2006

> I arrived in Egilsstaðir at three o'clock on a very grey Sunday afternoon, from the rainy southeast. It was snowing heavily on Öxi and the shortcut mountain road was only barely passable. Down in Skriðdalur the snow turned to sleet. In town I was greeted by a strange sight: small groups of people wandering the streets in the town centre, carrying plastic shopping bags. Most of these had a decidedly non-Icelandic look, their creased faces and comportment indicating years of enduring hard work and extremities of weather. I heard Portuguese and Chinese spoken in the streets.

> Local people later told me that on Sunday afternoon the bus comes down from the dam building site on the high plateau with those who want a break. But there is very little to be done in Egilsstaðir on a Sunday. The main supermarkets had been opened, though. Hence the visitors passed their time wandering between the shops, buying nothing much – after all Iceland is very expensive and the wages for these international construction workers

are quite low. The shopping bags contained some low-alcohol Icelandic beer perhaps – Egils pilsner. I was told that sales of Egils pilsner had quadrupled.

When the bus left, later that afternoon, Egilsstaðir reverted to its usual quiet self. There was no sign at all of this temporary invasion from globalized 'outer space'. So near, yet so far away.

* * *

All the projects associated with the power development scheme needed human energy, and lots of it: construction workers. A scattering of international migrant labourers had started coming to Iceland in the 1990s, but with Kárahnjúkar this trend speeded up dramatically.

In 2006, at the height of the construction period, some 1,500 people were at work at the smelter site, with a similar number at the dam site and other sites relating to the power station. Some efforts were made to recruit Icelandic workers, both from the region and from elsewhere in Iceland, but it was obvious from the start that the project needed a large component of foreign labour: about three out of every four workers were supposed to come from other countries.

The Italian-based construction conglomerate Impregilo had been the lowest bidder for the dam construction and tunnels, by a large margin. The firm had extensive experience from large-scale contracts in other parts of the world. Conditions at Kárahnjúkar turned out to be unusually tough, however. Warning lights started flashing in the Icelandic media. Some people alleged that Impregilo's performance in matters of worker's rights and environmental matters had been less than stellar, especially in their many projects in poorer countries. The company would in any case have to cut some corners, the sceptics argued, were it to make an acceptable profit for its shareholders.

Large labour camps were erected on the highland plateau. These were occupied by a multinational army of migrant labourers. The first batches of workers came mainly from Portugal and Italy, but many other nationalities arrived at the regional airport at Egilsstaðir, to be transported to the project site: Romanians, Croats, Lithuanians and many more. Neither the contractor nor the forces of nature treated the incomers particularly gently. Before long, news about a blatant disregard for wage agreements, bad conditions at the workers' camp and lack of attention to workplace safety at the Kárahnjúkar site became common fare in the Icelandic media. To be fair to Impregilo, these incidents just as often related to their Icelandic sub-contractors as to the Italian company itself. Come winter, the workers found themselves operating in an extremely unforgiving nature, where blizzards and biting frosts made life difficult at times in the flimsy labour camp, as well as in the workplace. The workers complained.

Eventually the construction company brought in a large group of Chinese workers, explaining to the media that the Chinese were particularly 'hardy' (Icel. *harðgerðir*) and thus fit to work in these difficult circumstances. The Icelandic unions suggested, however, that the 'hardiness' of the Chinese might also have to do with their willingness to accept working conditions and pay packages that workers from Southern or Eastern Europe would not be content with. For the entire duration of the project, suspicions and accusations surfaced constantly about non-compliance with official wage agreements, harsh treatment of workers, and so on. Accidents at the work sites were common, including several fatal accidents. To Icelanders, the Kárahnjúkar project was a rude awakening to the brave new world of globalized labour, mobile and docile – yet 'hardy', where the worker seemed to be less of a subject than an object: a mobile body, put to work in the production of a landscape of accumulation.

Things turned out somewhat differently down on the coast at the smelter site. Here, the main contractor was another global player, the American Bechtel Corporation – also a corporation with a long association with large-scale constructions around the world,[2] but with a very different corporate culture in matters relating to occupational health, and safety in particular. A large temporary camp was erected here as well, just outside the little town of Reyðarfjörður. The majority of the workers came from Poland. The construction company placed great emphasis on accident prevention and seemed to be able to steer clear of many of the problems with labour relations that had dogged the dam project.

Second Story: Moving Images[3]

From the newspaper Morgunblaðið

> It is enough to look at photographs of the land which is to be submerged to feel pain in the heart. This pain is not measurable, which may be irritating for the men with the measuring instruments. Without wanting to denigrate measurements, not everything is measurable. We have neither been able to measure the length of love nor the circumference of God. (Jökulsdóttir 2002, 38)

* * *

The opponents of the Kárahnjúkar project enrolled numerous and diverse arguments in their struggle. Many of these arguments were based on established lines of 'rational' academic reasoning. Several economists questioned the assessments that had been made of the project's profitability. They criticized the secrecy concerning true electricity prices in the agreement made between the publicly-owned National Power Company and Alcoa, and pointed to the many risks associated with a project of this magnitude. Social scientists expressed doubts about whether the industrial development would manage to stem out-migration from the settlements of East Iceland in the long run, given the many socio-cultural variables involved in that equation, aside from just the availability of jobs. Natural scientists pointed to the loss

---

2   As well as in other sectors: Bechtel was, for instance, an active player in much-maligned, World Bank-led moves to privatize water supplies in some Third World cities. The opponents of the Icelandic mega project invoked this in their bid to mobilize Icelandic resistance to the project. Bechtel is also among the leading contractors in the American 'reconstruction' of Iraq.

3   This story is also told in Benediktsson (2007), in the context of landscape aesthetics.

of unique sedimentary land forms and large tracts of well-vegetated land, which is in short supply on the highland plateau of Iceland.

But these arguments carried only limited weight. Many of them were met with equally 'rational' scientific counter-arguments, making it difficult for the lay person to decide upon the validity of each. From the outset, however, the battle also inevitably comprised aesthetic and ethical aspects, which spoke more directly to the sensibilities of the nation. Like all conflicts of this sort, the Kárahnjúkar conflict became intensely emotional, with protesters exploring and explicitly transgressing the limits of instrumental rationality. It became a clash of sometimes apparently incompatible reasoning.

Visual aesthetics played a key role in this regard. Images showing the varied beauty of the area to be submerged proved a particularly potent force for mobilizing resistance. This is interesting, because the area in question had until then been visited by very few people. Those who knew its landscapes best were farmers from the valley below, who had used the area as grazing commons for ages. Hardly any tourists had been there and the area had not been known for any spectacular 'sights' – even if the 200-metre-deep canyon gouged out by Jökulsá á Brú at Kárahnjúkar soon achieved such status as the project progressed. To the casual observer, reaching the end of the road at Kárahnjúkar and gazing southwards through the windscreen of her 4WD, the site of the reservoir seemed a rather unremarkable tract of land.

The work of several photographers went some way towards changing this perception, thus 're-enchanting' this landscape (see also Chapter 19 by Guneriussen). Landscape photography is well-established in Iceland. Glossy coffee-table presentations of Icelandic nature and landscapes are a staple of the tourism industry, often utilizing standard motifs and techniques of wide-angle photography (cf. Hafsteinsson 1994; Óladóttir 2005). In the Kárahnjúkar case, photographic representation took a somewhat more complex and interesting turn. Two photographers were especially notable. Jóhann Ísberg, a keen naturalist and photographer, set out to document most of the area photographically in a systematic manner. His photos were put up on a web site organized by a conservation NGO and were widely used to demonstrate the 'hidden' geological and biological diversity of the area. However, the professional photographer Ragnar Axelsson probably achieved the most poignant and influential pictorial intervention. Axelsson worked as a photojournalist for Iceland's most respected newspaper, *Morgunblaðið*, and had gained various distinctions for his work. In late 2002, the paper published his photographic essay in three parts, under the title *Landið sem hverfur* (The land that will disappear). Succinct captions, highlighting both the landscape and biological conditions, and what would be lost if the project went ahead, drove the message home. The photos were later exhibited in the Kringlan shopping mall, which ensured that more or less half the country's population saw them. They created a lot of interest and commentary. Some were even suspicious: the colours were certainly rigged, some people said; there simply could not be such colourful rocks and vibrant vegetation tucked away up there on the grey and stony highland plateau. Many members of the general public seemed to be genuinely moved by these images.

Both photographers sought to portray not only the usual fare of Icelandic landscape photography: dramatic scenery, deep canyons and mighty waterfalls. To be sure, such photographs were there, as all these features are to be found in the area. But the photographers also directed their attention to beauty of a more delicate nature: to small-scale artworks of rocks, water, plants, birds and animals. Yet above all, the viewers' response was a result of their knowledge of the planned fate of these small wonders. The photographs showed a nature that was to be mutilated, their visual language exposing the one-dimensional ideological premises of the hydropower project. A philosopher and cultural critic observed that Axelsson's photos could be regarded as

> a reflection on a world that was – nature already sentenced to death in the name of interests which nobody is totally certain are the real interests in the long run. …The decision has been made, but the sacrifice is nevertheless obvious: a sacrifice of life. It therefore looks as if those who make the decision – those who speak for rationality, those who speak for industry and economy – have to make a leap of faith in the end: carry out the sacrifice with the blindness of one who really does not know what the future holds, yet puts trust in one's religious conviction. (Ólafsson 2003, 80–81)

To many people, one photograph, or rather one motif, proved particularly moving. This showed a distinctive rock formation, situated right on the bank of Jökulsá á Brú, at the bottom of the future reservoir behind the dam at Kárahnjúkar. Located in a spot very seldom visited by human travellers, this rock formation had been almost unknown previously except to a few farmers from the valley below. Axelsson's picture effectively highlighted the anthropomorphic features of the rock (see Figure 17.1).

Figure 17.1   The disappearing land

Source: photo by Ragnar Axelsson.

The photo became a defining image for the place to be 'sacrificed'. Conservationists found in this visually arresting figure a potent iconic symbol: a silent spokesperson, who turned out to be particularly effective in conveying the message that the Kárahnjúkar project was an affront against nature and landscapes. The anthropomorphic stone figure who 'stood guard' on the river bank invested the landscape with a moral purpose. Was this simply photographic propaganda? Indeed, a critical deconstruction is both possible and appropriate. The power of the photographs was linked with references to natural history, mythology and nationalism. A particular place was being produced and animated through their conscious use. But this had been achieved, too, through the stories of engineering feats and economic bonanzas that for the most part had dictated the terms of the debate.

The stone figure by the river was never 'drowned' in the reservoir. *Morgunblaðið* told its readers in May 2005 that it had ceased to stand guard and had disappeared altogether, leaving only a small pile of rocks behind. Sometime during the previous winter the river itself had ended its existence, saving it from the indignity of submersion.

Third Story: Mobile Protesters

Statement from the website 'Saving Iceland'

> We have gathered to protest the continuing devastation of global ecology in the interest of corporate profits. We have come here to tip the balance of a struggle portrayed to be national, while actually being much larger: from the Narmada Dams in India, to the proposed Ilisu Dam in Turkey, the story is one of big business and oppressive government. The struggle to save our planet, like the struggle against inhumanity, is global, so we have to be too. We're here to prevent the Kárahnjúkar Dam project from destroying Western Europe's last great wilderness...
>
> ...Across the world, people are coming together to oppose the blatant lies, corruption and oppression generated by corporations and governments alike. In this spirit, we are asking that all those opposing the Kárahnjúkar Dam organize or partake in solidarity actions globally or locally. (Saving Iceland 2007)

* * *

Environmental activism has long since become a mobile, international force. International environmental organizations and individuals soon started to move when the Kárahnjúkar project got off the ground. During the summers of 2005 and 2006, Icelanders saw for the first time the spectacle of confrontational activist protests,[4] which many other nations had experienced much earlier. A group of mostly foreign protesters set up camp in the vicinity of the dam building site and pulled a series of stunts designed to interrupt the construction work, and thus attract media attention to the cause. They entered areas of restricted access, chained their own bodies to bulldozers and trucks, and performed various other acts to underline their message. The organizers of all this were *Saving Iceland*, a group whose existence seemed to be largely virtual: an example of a form of activism increasingly engendered by the Internet since at least the anti-globalization Battle of Seattle in 1999. As the quotation above indicates, Kárahnjúkar did indeed seem to have become a symbolic

---

4   This is not quite true: an incident of this sort had occurred in 1986, when two whaling boats were sunk in the harbour of Reykjavík by members of the Sea Shepherd Conservation Society. By comparison, the Kárahnjúkar protesters were not very militant at all.

site of resistance to global capitalism, with a significance well beyond Iceland and beyond nature protection *per se*. It was all resolutely non-violent, though.

Not that this was the first instance of protest. In fact, opponents of the project had voiced their concerns loudly and repeatedly right from the start. Numerous protest meetings and rallies had been held, both in Reykjavík and elsewhere. Most of these had been rather polite, certainly not transgressing legal limits of any sort. This seemed to be different. The local chief of police used stern words to express his indignation and in 2006 the police eventually closed down the protest by a show of force that seemed to many observers to be out of all proportion, given the repeatedly declared non-violence of the group.

The direct and confrontational methods of these protesters were actually met with a rather negative sentiment by many 'ordinary' Icelanders, even many of those who sympathized with their cause. One form of 'bodily' protest – a very different one – seemed to be more successful in a way. This consisted of organized hiking tours through 'the disappearing land' north of Vatnajökull. The two female organizers, a yoga teacher and tour guide, and a visual artist, both of whom had worked for many years as hiking tour guides, said that they wanted to give the nation the opportunity to experience with its own senses the beauty of the nature in question: the rock formations and wildflowers that would be submerged; the numerous waterfalls that would be either drowned in the reservoir or emptied of water. They stressed

**Figure 17.2   A moment of contemplation during a hiking tour through
                'the disappearing land'**

Source: by Hjörtur Hannesson.

the spiritual content of such a journey, where conventional hiking would be mixed with yoga sessions and communing with nature (Figure 17.2). This was not to be a highland hike just like any other, but a hike that would have a transformative effect. The trips, which took several days, ended with the participants standing in front of the construction site for some minutes in a silent protest.

The first of these tours took place in 2003, when the building of the dam had only just started. Not many people took the idea seriously, and the dam proponents dismissed it as irrational New Age nonsense. But gradually, as the dam building project progressed, more and more people felt this to be a meaningful act of solidarity with the nature that was to be obliterated or altered. Word spread and the tours proved increasingly popular. Most of those who went had of course made up their minds already. But many participants stated afterwards that being physically present in this 'doomed' place and witnessing its geological and biological wonders, large and small, had revealed to them a previously hidden world of radically different values, priorities – and possibilities that had now been foreclosed.

Conclusion: Stories of Moving Places

> If space is [...] a simultaneity of stories-so-far, then places are collections of those stories, articulations within the wider power-geometries of space. Their character will be a product of these intersections within that wider setting, and of what is made of them. And, too, of the non-meetings-up, the disconnections, and the relations not established, the exclusions. All this contributes to the specificity of place. (Massey 2005, 130)

The preceding account has sought to provide an insight, however selective, into the ongoing reconfiguration of places that is occurring in the North. No less than places elsewhere, northern places are constantly being re-articulated – or produced anew – at the intersection of economic, social, cultural, and indeed natural forces, as other chapters in this book also show. In short, they may be understood not so much as stable, timeless containers as the products of relational enactments, or assemblages (cf. Braun 2006).

'Nature' has always been a strong place-making force in the Nordic periphery. Numerous communities have based their existence on 'natural resources', whether fish, timber, minerals or other materials desired by the wider world and produced within the framework of national economies. Yet 'natural landscapes' have arguably only recently emerged as a major, explicit *political* force, in and through relations that compose often contradictory ideological landscapes.

The highland plateau of Iceland is a particularly clear example of this trend. Traditionally imagined largely in terms of folklore and farming – as an ambivalent place of apprehension and utility, respectively – it is now caught up in much more complex and highly politicized processes of place-making, which draw upon ideologies and emotions with more or less global currency (see also Chapter 16 by Lehtinen, Chapter 18 by Kraft and Chapter 19 by Guneriussen).

The mountains of Kárahnjúkar and their surroundings comprise a 'moving place' in many senses of the word (cf. Massey 2006). Seen on a geological timescale, at least, its 'natural' landscape is quite obviously dynamic and mobile. But what the stories told in this chapter have also shown is the capacity of that 'natural landscape' to mobilize diverse actors in its own right (cf. Castree and Macmillan 2001; see also Chapter 13 by Jóhannesson and Bærenholdt), which in turn ceaselessly reworks and reconstitutes it – turning it into a place of 'socionature' (Braun 2006). We found mobile capital in search of a place to anchor itself to, at least for a while. There were itinerant construction workers looking for a place to underwrite the economic survival of themselves and their families back 'home'. Last but not least, the place itself, both through direct sensory contact and the images of it that were circulated, moved a large and diverse group of concerned individuals to action against the power project. Even if a view of place as a relational enactment is adopted, place must still be recognized as an enormously important locus of affect and emotion.

Acknowledgements

Parts of this chapter are based on fieldwork conducted in East Iceland in 2006 for the project 'Place Reinvention in the Nordic Periphery', funded by Nordregio (Nyseth and Granås 2007). The Icelandic part of the project did not specifically focus on the Kárahnjúkar project, but on the small towns in East Iceland (cf. Benediktsson and Suopajärvi 2007; Benediktsson and Aho 2007). The fieldwork assistance of Bryndís Zoëga is acknowledged. The author would also like to thank Ragnar Axelsson and Hjörtur Hannesson for permission to publish the photographs accompanying the text.

References

Appadurai, A. (ed.) (1986), *The Social Life of Things: Commodities in Cultural Perspective* (Cambridge: Cambridge University Press).
Benediktsson, K. (2007), 'Scenophobia', Geography and the Aesthetic Politics of Landscape, *Geografiska Annaler* B 89:3, 203–17.
Benediktsson, K. and Aho, S. (2007), 'Concrete Messages: Material Expressions of Place Reinvention', in Nyseth and Granås (eds).
Benediktsson, K. and Suopajärvi, L. (2007), 'Industrious Cultures? The Uneasy Relationship between an Industrial Order and a "Second Modernity"', in Nyseth and Granås (eds).
Braun, B. (2006), 'Environmental Issues: Global Natures in the Space of Assemblage', *Progress in Human Geography* 30:5, 644–54.
Braun, B. and Castree, N. (eds) (1998), *Remaking Reality: Nature at the Millennium* (London and New York: Routledge).
Castree, N. and Braun, B. (eds) (2001), *Social Nature: Theory, Practice and Politics* (Oxford: Blackwell).
Castree, N. and MacMillan, T. (2001), 'Dissolving Dualism: Actor-networks and the Reimagination of Nature', in Castree and Braun (eds).
Friðriksson, G. (1997–2000), *Einar Benediktsson: Ævisaga* (Reykjavík: Iðunn).

Hafsteinsson, S.B. (1994), 'Fjallmyndin: Sjónarhorn íslenskra landslagsljósmynda', in Magnúsdóttir and Bragason (eds).

Hálfdanarson, G. (1999), 'Hver á sér fegra föðurland: Staða náttúrunnar í íslenskri þjóðarvitund', *Skírnir* 173, 304–36.

Jamison, A. and Østby, P. (eds) (1997), *Public Participation and Sustainable Development: Comparing European Experiences* (Aalborg: Aalborg Universitetsforlag).

Jónsson, Ö.D. (1997), 'The Uncompromising Ally: Environmental Policy in Iceland', in Jamison and Østby (eds).

Jökulsdóttir, E. (2002), 'Um ástina og aðrar tilfinningar', *Morgunblaðið* 1 November 2002, 38.

Landsvirkjun (2002), Annáll: frá Kárahnjúkum 2004 til Bessastaðaár 1946. *Kárahnjúkavirkjun*, from http://www.karahnjukar.is/.

—— (2007), *Kárahnjúkavirkjun*, from http://www.karahnjukar.is/.

Macnaghten, P. and Urry, J. (1998), *Contested Natures* (London: Sage Publications).

Magnúsdóttir, E.B. and Bragason, Ú. (eds) (1994), *Ímynd Íslands: Ráðstefna um miðlun íslenskrar sögu og mennigar erlendis* (Reykjavík: Stofnun Sigurður Nordals).

Massey, D. (2005), *For Space* (London: Sage).

—— (2006), 'Landscape as a Provocation: Reflections on Moving Mountains', *Journal of Material Culture* 11:1/2, 33–48.

Nyseth, T. and Granås, B. (eds) (2007), *Place Reinvention in the North. Dynamics and Governance Perspectives* (Stockholm: Nordregio).

Óladóttir, H. (2005), *Með kveðju frá Íslandi! Íslensk landslagspóstkort, myndefni þeirra og sala* (Unpublished MS-dissertation, University of Iceland, Reykjavík).

Ólafsson, J. (2003), Myndir áróðursins, *Ritið* 3:1, 79–92.

Ólafsson, K. (ed.) (2006), *Áfangaskýrsla I – Stöðulýsing og upphaf framkvæmda á Austurlandi* (Akureyri: Byggðarannsóknastofnun Íslands).

Saving Iceland (2007), from http://www.savingiceland.org/.

Skúlason, J.B. and Hayter, R. (1998), 'Industrial Location as a Bargain: Iceland and the Aluminium Multinationals 1962–1994', *Geografiska Annaler* B 80:1, 29–48.

Staðarvalsnefnd um iðnrekstur (1983), *Staðarval fyrir orkufrekan iðnað: Forval* (Reykjavík: Iðnaðarráðuneytið).

Whatmore, S. (2002), *Hybrid Geographies: Natures, Cultures, Spaces* (London: Sage).

Chapter 18

Place-making through Mega-events

Siv Ellen Kraft

We want to thank the people of Tromsø for the courage to recognize our common humanity and all that it entails. In recognition of this leadership we have made Tromsø the first ambassador city for 46664. Many years ago, I said that my long walk has not yet ended. As we stand here tonight, I gain great comfort in the knowledge that we are not alone on this journey. In taking on this challenge you have become part of us and so you are now all Africans.

Nelson Mandela

On 11 June 2005, a Nelson Mandela solidarity concert took place in Tromsø, the regional centre of Northern Norway. Dedicated to the fight against HIV/AIDS, the event saw the gathering of 18,000 people – musicians, journalists, celebrities and politicians from around the world, and locals of all possible sorts and ages. The world came to Tromsø on the night of the concert, and Tromsø introduced itself to the world. This was a mega-event in a small place, a place moreover that has traditionally been relegated to the northern periphery and Norway's backyard. The concert provided alternative images – literally and metaphorically. Local newspapers reported on a magical night, on a demarcation line in the history of Tromsø, and of 'little Tromsø's' contribution to worldwide changes. They also spoke of the images sent out into 'the world' – of what the international press would see and report.

Place-making and mobility come together in the stories to be analysed in this chapter. The importance of the event depended upon people and images travelling, physically and via the press. It was this same measure of importance that probably fuelled local enthusiasm. It certainly fuelled enthusiasm among local journalists, who constitute my main sources in this article. The concert was a principal topic in Tromsø's two main newspapers during the summer of 2005. By focusing in detail on the event, journalists helped to shape people's experiences at the concert, as well as consolidating memories and linking them to revised and more positive narratives of the place and its meaning.

My theoretical approach is informed by perspectives developed within media studies and religious studies, including theories concerned with mega-events, ritual and civil religion. I am interested in global scripts behind place-making creativity, and with ritual as a means of experiencing and dramatizing stories in the making. I argue, moreover, that the scripts of mega-events and civil religion overlap to a large extent with regard to place-making issues. Mega-events are patterned upon a global script which emphasizes the uniqueness of the local place (Urry 2003). What I refer to as the emergent features of a global civil religion adds a sacred or semi-sacred dimension to such uniqueness, and at the same time connects it to a global level of pan-human needs, issues and promise.

This chapter commences with a discussion of ritual theory and the relationship between the global and the local – between mega-events as civil ritual and the emergent features of a global civil religion. This is followed by introductory comments on the local setting – the 'city'[1] of Tromsø and issues relevant to the following analysis. Finally, my analysis of local newspapers revolves around two overarching orientations in their coverage of the concert: on the one hand an emphasis on uniqueness and difference, and on the other hand a focus on pan-human ideals, values and characteristics.[2]

Ritualizing the Place – Public Rituals and Ritual Effervescence

My ritual approach draws upon elements from performance theory, a broad field of overlapping approaches, mainly concerned with the question of how rituals work: what their power consists of, and how emotive, physical and sensual aspects contribute to this power (Bell 1997, 73).[3] Moreover, I shall attempt to develop established categories of civil rituals with regard to globalization. No longer exclusive to the national context, mega-events may also 'belong' to the local place and to regional identity creation.

Ritual performance

Performance theorists accept the Durkheimean notion of ritual as characterized by effervescence, and such effervescence as critical to the moulding of individuals and societies. However, effervescence is not related to a general rule of functionalism. Rituals may be more or less successful, more or less functional, and may play a more or less active role in constructing reality and contributing to changes in values and ideas.

Among the theoretical concepts developed to approach such issues, what I refer to in the following as 'framing', 'display', 'public reflection' and 'multisensory appeal' are basic. Framing, first, has to do with the setting apart of certain activities, thus allowing for an alternative interpretive framework 'within which to understand other subsequent or simultaneous acts or messages' (Bell 1997, 74). In the case of the concert, local newspapers contributed to these processes in important ways. In addition, the week preceding the concert saw a host of related activities, an unusually high number of tourists and visitors and an extensive reorganizing of the town surface, including huge Mandela posters, concert programmes, ticket booths and souvenirs.

---

1   Technically, Tromsø is not a city, but it is referred to as a city by the municipality and by local travel companies (see www.tromso.kommune.no and www.destinasjontromso.no).

2   This chapter is a revised and shortened version of a previous publication (Kraft 2006a). The initial, theoretical part of the chapter overlaps with the previous version.

3   The performance approach to ritual studies dates back to the 1970s and the coming together of several different contributions, including Victor Turner's description of ritual as 'social drama', Austin's theory of performative utterances, Kenneth Burke's discussion of dramatism and Erving Goffman's work (Bell 1997, 73).

Thus prepared for an extraordinary event, 18,000 people gathered in Tromsdalen on the evening of the concert. A large gathering by any standards, this was huge for a town of 62,000 inhabitants. This was a mega-event in a small place, a place that had never hosted anything of this magnitude and whose population was very conscious of the world watching. In the language of performance theory, the concert was an occasion for ritual display, and ritual display may be expected to serve at least two tasks. On the one hand it contributes to a sense of awe and of being overwhelmed, in this case relating to size, spectacle and the presence of Mandela. On the other hand, display – as a way of doing (rather than merely saying) – demonstrates the power, stories or messages presented, thereby providing them with factual status (Rappaport 1999, 56).

Public reflection may be related to the otherness of liminal times or places, and the break from routine that follows on from the arrival of the extraordinary. In this way, rituals may help to create a distance necessary for reflection to take place and 'live' material for the mind to ponder upon. The presence of the press further encourages such a distancing. Newspapers and television quite literally enable crowds to become audiences of themselves, to view themselves from outside and – in the case of mega-events – to do so in the company of distant viewers.

Finally, 'multisensory appeal' refers to the potential for ritual to communicate on several levels simultaneously. In this instance, the combination of the crowd, music and outdoor settings was no doubt important. After weeks of heavy rain, this was a mild and sunny night, perfect for an outdoor concert and likely – in the context of visitors and worldwide broadcasting – to bring out a pride of place among the locals. Music, dance and concert script, meanwhile, literally co-ordinate and connect a crowd, and may thereby contribute to feelings of wholeness and belonging (Bell 1997).

Mega-events as civil rituals

Specific historical settings tend to produce similar patterns and types of ritualization. In late modern Western societies a dual tendency has emerged, with an extensive privatization of ritual on the one hand, and on the other, a vast body of vaguely religious 'civil rituals', such as pledging allegiance to the flag, the swearing-in of political leaders and various types of festivals (Bell 1997). Civil rituals typically have little of the traditionalism and invariance typical of Western rituals in the near past. They emphasize moral and ethical commands over ritual duties, and their sacred symbolism tends to be implicit or ambiguous (Bell 1997, 201). As such, they provide room for the heterogeneous 'congregations' of late modern society and religion placed outside the boundaries of organized religion. As such, they provide room for integrative functions. In civil religious rituals – to the extent that these are successful – participants enact and celebrate their identity as a group and the values they stand for.

Robert Bellah's *Civil Religion in America* (1967) is the seminal study in connection with civil rituals. In line with his work, most contributors have related civil rituals to the national context, with civil rituals as expressions of civil religions, which in turn consist of important national ideas and values. Mega-events have similarly been related to the national framework as 'important to the story of a country, a people, a nation' (Roche 2000, 6).

In an article about Princess Diana, Roland Boer suggests an extension to this perspective. Civil religion has more recently moved on to a global level, he argues. As a global phenomenon, civil religion 'has been creeping, perhaps a little more slowly than other dimensions of culture, into the global arena, particularly through a global media that is able to transmit and present iconic figures to a world-wide audience' (Boer 1997, 83). Boer mentions Mandela, Mother Theresa and Princess Diana as examples of global icons, but does not further elaborate upon the globalization of civil religion, and leaves out the possibility of local variation.

Adding to his insights, one might argue that the semi-sacred status of such figures speaks to the semi-sacred values of the world order – to core modern values such as liberty, equality and fraternity (the ideals of the French Revolution) and to more recent additions, such as pluralism and environmentalism. These same values are crucial to global institutions such as the UN, and mega-events are venues for expressing them. Mega-events frequently include a visionary and moral dimension, and those which do invariably relate to core modern values. The alternative is not realistic. It is hard to imagine an Olympic ceremony constructed on the ideals of a homogenous society, the defence of national boundaries and the racial purity of the host population, or one that speaks explicitly against environmentalism, gender equality or peace.

A fear of cultural homogenization is crucial to this emerging global civil religion. UN strategies, particularly with regard to indigenous people, take it as a premise that cultural diversity is natural and positive, and that the uniqueness of 'cultures' should be protected (Lechner and Boli 2005). On one level, then, the emerging global civil religion may be expected to encourage local diversity, and such diversity may not necessarily be restricted to the national level. Rather, the diminishing status and power of nation-states may encourage alternative routes to world recognition. This may be particularly the case with regard to places on the periphery that have traditionally felt left out of the national project.

The Local Setting – Place Branding and a Sense of Place

As an example of places on the periphery, Northern Norway has traditionally been constructed according to a north-south axis, with 'south' as the centre of power and decision-making, and 'north' as a suppressed and exploited backyard. Related to this perspective of subordination and victimization, the people of the north have been imagined through a primitivist discourse, in contrast to a presumed modern, Western or, in this case, Norwegian identity (Geertz 2004). Still common in mass media portraits of Northern Norwegians, primitivism emphasizes smallness, periphery, nature and tradition. Northern Norwegians are nature-people: living in close contact with nature and moulded by rough, dramatic and spectacular landscapes. In order to survive in this situation they have developed strength and roughness, willpower and a sense of humour. Like primitives of all kinds, they live in close-knit villages, characterized by solidarity and community, and in line with these pre-modern conditions people tend to be simple and somewhat naive. Ethnic diversity has further contributed to the image of Northern Norway as different and only ambiguously Norwegian.

In recent years, these stereotypes have been confronted and questioned. Critics have called for 'new stories', more apt to the present situation of Northern Norwegians, and particularly to the 'cosmopolitan' population of Tromsø.[4] The regional centre of Northern Norway, Tromsø has 62,000 inhabitants. It is the largest Norwegian town north of the Arctic Circle and in terms of educational and cultural facilities it is by far the most developed. The only university in Northern Norway is located here, along with several regional institutions. In addition, the northern area has been selected by the Norwegian Parliament as a main strategic concern with regard to research, environmental measures and natural resources.

An interest in hosting mega-events speaks to this recent optimism and the related search for 'new stories'.[5] Two different mega-events have been proposed and debated in Tromsø over the last couple of years. Prior to the Mandela concert, a local committee developed a proposal for the 2014 Winter Olympics in Tromsø. Although rejected by the Norwegian government, the proposal paved the way for the Mandela concert in at least two important ways. Firstly, the proposal appears to have raised the level of ambition and sparked interest in mega-events and the town's potential for hosting them. Secondly, the local planning committee provided a new, coherent and extremely ambitious concept of Tromsø and its potential. Based upon the overarching ideals of Olympism,[6] the proposal offered a coherent story of local uniqueness and global commitment – with Tromsø as a town of peace, pluralism, justice, initiative and 'Arctic magic'. Newspaper coverage of the Mandela concert drew extensively upon the ideological scenarios presented here. By actually taking place, the concert provided life and solidity to visionary creativity. Thus, we have an element of dialogue between the Olympic proposal and the Mandela concert – between myth-making and ritual display.[7]

---

4   Since the end of the 1990s, the issue of place identity has been an important topic in local newspaper debates. See also Paulgaard 2006.

5   Maurice Roche mentions South Africa as an example of a country whose government attempted to change the political identity of the nation through an Olympic Games arrangement (Roche 2000). Roche also considers Mandela's willingness to support and participate in mega-events (including South Africa's bids for the 2006 World Cup and the 2008 Olympics) as proof of their importance: 'the fact that a politician of his stature would choose to continue to serve his vision of his nation by being involved in such bids speaks eloquently for the importance of mega-events for nations in terms of their self-image and place in the world society' (2000, Preface).

6   The modern Olympic Games were revived by a Frenchman Pierre Frédy, Baron de Coubertin, in Athens in 1896. His expressed goal was to unite mankind through an international festival dedicated to athletic greatness (Chaffer and Smith 2000, 2).

7   This dialogue, moreover, has not yet ended. In March 2007, the local committee sent a new proposal for the Olympic Games 2018 and this proposal is still a major topic in local newspapers. Recent surveys indicate massive local support, but the proposal has also proved controversial. An organization dedicated to the fight against the Olympic Games in Tromsø was established during the first round, and this is still active.

The Mandela Concert – Themes and Issues

Local and regional newspapers are important public forums in Norway. In Tromsø, these forums – particularly *Nordlys* – have taken an editorial stance in favour of new and more ambitious place stories. Both *Nordlys* (a regional newspaper) and *Tromsø* (local) have also embraced the two mega-event initiatives, and have explicitly and extensively related them to issues of place-making. What Stewart Hoover has referred to as the 'almost priestly role' of journalists is clearly at work in these cases. By priestly role, he implies a particular level of participation, characterized by a suspension of 'their normal critical stance' and a treatment of 'their subject with respect, even awe' (Hoover 2006, 254). In the case of the Mandela concert, the metaphor of the priest is relevant with regard to both the journalists' treatment of their subject and the more general religious or quasi-religious dimension granted to it.

Local uniqueness – a small place on the edge of civilization

In *Media Events: The Live Broadcasting of History* (1992), Daniel Dayan and Elihu Katz suggest three story forms or scripts among the narrative possibilities of the genre. Among these, the category of 'conquest' refers to 'giant leaps for mankind' – rare events, as a result of which life will never be the same. Indicative of the tone of newspaper coverage of the concert, journalists agreed upon a narrative of triumph and of great change, for Tromsø and for the world. Journalists spoke of a 'historical event', of 'witnessing history', of a demarcation line in the history of Tromsø and of Tromsø being placed on the world map. 'We have moved the world', *Nordlys* claimed on 13 June; journalist Linda Vaeng Sæbbe later added that 'We have made the impossible possible. It is too early to say what kind of forces this in turn may release. But it will be grand and important, of that I am certain' (*Nordlys* 16 June 2005).

Tromsø's *mission impossible* was also a triumph *against all odds*. Broadly in line with the traditional north-south perspective, journalists spoke of a 'wild idea', of 'doing the impossible', and of opposition from the south – in this case represented by a lack of support from the Norwegian government. In addition, co-operation with international organizations proved frustrating and difficult. Recalling initial meetings with leaders of the Nelson Mandela Foundation, a member of the local organizing committee noted that they had not been taken seriously: 'a tiny dot in a remote place in a remote country', Tromsø was not the best starting-point for a gathering of international stars and worldwide attention (Hansen, *Nordlys* 18 June 2005). On the night of the concert, however, Tromsø proved itself to its inhabitants and to the world. As one journalist put it, in response to his own question 'what will people out there see on their TV screens?'

> A small town near the North Pole, which has thrown itself into Mandela's fight, a place where the sun never sets, where people join to care for others, and where fellowship is so strong that even bad weather must give in when people want it enough. (*Nordlys* 13 June 2005)

Several themes come together in this citation, all of them prominent in the newspaper coverage. Firstly, the success of the concert was related to the collective efforts of the people of Tromsø, to what several commentators referred to as the 'Tromsø spirit'. Journalists told the story of organizers' efforts, but emphasis was placed on what 'we' had achieved, made happen and experienced.

Secondly, frequent reference was made to size and setting. In contrast to the more typical setting of mega-events, the uniqueness of this one was partly related to the fact of being a small town in the northern periphery. Both organizers and journalists emphasized the extreme and unusual features of the place, and in both cases 'Arctic' and 'magic' served to underscore and enchant these features. More or less everything was described as Arctic and magic – the landscape, the evening, the atmosphere, Mandela, his performance, the success, and – last but not least – the weather: after weeks of cold and rainy weather, the sky cleared on the day of the concert; rumour has it that a local shaman was behind this deed!

'Arctic' and 'magic' overlap and intersect in these stories. Both belong to the traditional vocabulary of regional images, and both – due to changes in late modern culture – have been granted new and positive meaning. A brief historical background may illuminate these changes. We shall start with 'the Arctic', and then move on to 'magic'. During the course of the twentieth century, sociologists John Urry and Phil Macnaghten have argued, 'the division between nature and society increasingly came to take a spatial form, with society in and at the centre and nature as the "other" pushed out to the margins' (Macnaghten and Urry 1998, 13). To be defined as 'nature' was to be defined as passive, as non-agent and non-subject. 'Arctic' is an extreme type of nature, according to this perspective, and it is an archetypical nature – 'more natural' than the rest of Norway and core places of global geography. In the twenty-first century, this position has been recast as an opportunity for uniqueness – as a resource to be employed by 'the other' as images of itself. The centres (of culture) have become increasingly linked to man-made risk, to mass-produced conformity and to superficial lifestyles. By contrast, 'nature' has come to signify origin, authenticity, depth and even wisdom.[8] 'Nature people' are connected to the same qualities, and are at the same time granted responsibility for their preservation – for not giving in to the homogenizing forces of globalization, and for withstanding the standardization of technology and modernization.

The history of magic has a similar structure. The connection between magic and the Sámi dates back at least to the writing of Old Norse literature (Mathisen 2003). Throughout this period, magic has been mainly a discourse of otherness. The assumed magical practices of the Sámi were considered dangerous to their surroundings, and at the same time a sign of primitiveness. In recent years, two changes appear to have taken place in this discourse. Firstly, 'fear' has been replaced

---

8   Sociologist Kevin Hetherington, in an interesting study of new social movements, comments upon this point. His main argument is that marginality and periphery have become repositories of truth and wisdom (Hetherington 1997). From this perspective, everything that dominant culture has rejected, suppressed or displaced is considered interesting. Historian of religion Peter Beyer makes a similar point in a discussion of so-called 'nature religions' (Beyer 1998).

by fascination – at least in public discourse and the types of place branding discussed here. Secondly, and again primarily in relation to communications with the outside world, the field of magic has been extended to entire landscapes and populations of Northern Norway (see Kraft 2006b). What remain are the traditional references to ability and extraordinary potential. Magic, as used in the context of this concert, is the practice of doing the impossible and making sure that 'miracles' pull through. It has to do with unbelievable images and with impressions and experiences too grand to be captured by language.

Northern Norwegians are still 'nature people', according to these recent revisions, and 'the North' is still pluralistic and heterogeneous. However, both concepts have been granted new and more positive content. In their coverage of the concert, newspapers spoke proudly of the 101 nationalities represented in Tromsø, highlighting in particular the Sámi. Meanwhile, in a popular television talk show Gerard Heiberg, leader of the Norwegian International Olympic Committee (IOC), explicitly referred to Northern Norwegians as 'nature people', adding that '*therefore* their hosting of the Olympic Games would be unique'.[9] Local newspaper accounts never explicitly used the concept 'nature people', but revolved around similar interpretations of fantastic potential. Nor was this potential limited to the provincial needs of the local community. The uniqueness of Tromsø, as we shall see later in more detail, consisted partly of its willingness and ability to take on global responsibility: to care for others – in this instance a pandemic that threatens all, but the consequences of which have been particularly acute among African people.

As an identity strategy, the patterns described here are not unique. Rather, the last few decades have witnessed a shift from attempts to adjust to the national mainstream towards articulations of difference and particularity. Folklorist Torunn Selberg considers the Forest Finns as an example of this tendency, and the use of so-called heritage festivals as a widespread 'means' (Selberg 2006). The Mandela concert belongs to a similar trend, but in this case heritage is a silent dimension – a backdrop that adds meaning to new formulations, stories and symbols, but which is never explicitly articulated and discussed. Emphasis, in the Mandela concert as in the Olympic proposal, is placed upon the presence and the future – on what 'we are', 'might be' and 'are becoming'.

Global and universal – you are now all Africans

> The highlight [...] was of course Mandela's speech to the world from the stage in Tromsø. There and then we were one world. Little Tromsø and big Mandela.
>
> Amundsen, *Nordlys* 13 June 2005

If 'Arctic magic' spoke to place uniqueness, then Mandela's presence signified global connections and humanitarian ideals. Rather than undermining place uniqueness, this wedding underlined the modern features of place and people, including their ability and willingness to 'act globally'. Situated under the northern midnight sun,

---

9   The talk show, called *Først og sist* (First and Last) is presented by Fredrik Skavlan and has become extremely popular over the past couple of years.

Mandela *was* the bridge between particularity and universalism. Uniquely positioned to establish such connections, Mandela not only has an extraordinary political reputation. He is, in the words of sociologist John Urry, 'a signifier of the newly emergent global order' (Urry 2003, 80). Along with the Olympic flag, rainforests and Mother Theresa, Mandela 'reflects and performs a global imagined community uniting different peoples, genders and generation' (Urry 2000, 80). This is not the 'bad' globalization of cultural Disneyfication and crude capitalism, but the 'good' globalization of core modern values. More than any other living statesman, Mandela embodies such values and, like that of Tromsø, his is a voice from the margins.

The magic of the event was mainly linked to the position and presence of Mandela:[10] this was mega and magical because of Mandela. Journalists granted minimal space to the musical dimension; they reported on an encounter with Mandela, and they did so in a language of religious effervescence. Mandela transcends the level of political statesmanship in these stories. Journalists spoke of 'moments of magic' during his speech. The silence during Mandela's speech was not only total, but 'devout', with the hair on 18,000 necks reported to have risen uniformly (*Nordlys* 13 June 2005). The man himself was 'magically' transformed before their eyes, from being an old, tired man – before and after the speech. To some of the newspaper commentators, Mandela was even a Christ-like figure. A feature article in *Nordlys* relates the 'magic of the concert' to the presence of 'the greatest human being ever to visit the town' (*Nordlys* 21 June 2005). The contributor notes that when

> with an inner power similar to no other human being in the world he called us all Africans, only Jesus appearing in Tromsdalen could have been bigger. I remember thinking: why does it have to be 2,000 years between every time the earth manages to produce a human being of such calibre. (ibid.)

Descriptions of Mandela as a divine figure are not unique to Tromsø. Upon Mandela's release from prison in 1990, Reverend Jesse Jackson described him as 'a Christ-like figure' who had 'suffered his way into power', adding that 'now that the stone has been rolled away' the world beheld an apocalyptic 'second coming'. During a visit to the US later that year, Mandela was publicly hailed as 'the most sainted man of our times', as a modern-day Moses (according to the then New York Mayor), as an African pope (according to Reverend Jim Holley) and, according to *Time* magazine 'a hero, a man, like those described by author Joseph Campbell, who has emerged from a symbolic grave reborn, made great and filled with creative power' (Chidester 1995, 291).

Basically in line with such interpretations, newspaper coverage of the Mandela concert added a link between 'us and him' – Mandela and his Northern Norwegian

---

10 Most people would probably agree that the list of performers was disappointing. Rumours of Bono attending did not materialize and the list of musical celebrities was limited to fading stars from the 1980s, African artists who are fairly unknown in Northern Norway, and local bands. This aspect of the event was the target of critical comment during the planning stages, but was more or less ignored in its aftermath. Having started out as an ambiguous combination of solidarity and musical event, the concert became situated unequivocally along the lines of the former – as a meeting with Mandela and a way of supporting his causes.

228 Mobility and Place

congregation. This was on one level a *rite de passage* – involving the transition to a new era in the history of Tromsø. Indicative of this theme, sacred symbolism tended to cluster around Mandela's speech, thus indicating the entrance of a liminal stage. In addition, some of the newspaper comments pointed in the direction of *communitas*, the temporary dissolution of distinctions typical of ritual liminality. The Chief Editor of *Nordlys* noted of the atmosphere during Mandela's speech that 'suddenly there was not much difference between small and large, between Crown Princess Mette-Marit and *Jørgen Hattemaker*'[11] (13 June 2005). Interestingly, the example used was the crown princess, rather than Mandela. Inter-structural time is not bereft of order or distinctions. Paralleling the dissolution of *most* distinctions, the remaining ones are further highlighted: in this case, the distinction between Mandela and the crowd, the leader of the ritual and ritual novices. True to the structure of rites of transition, moreover, ritual novices appeared to be changed by the process. Using an unusually explicit image to describe such a change, the editor of *Nordlys* described the concert departure of 18,000 people, walking 'in a powerful, kind and reflective procession out from the "dump", over the bridge and towards the sun' (Amundsen 18 June 2005). This was on one level a literal description: the concert was arranged at a former garbage dump, known to locals as 'the dump'; the bridge connecting the concert area with Tromsø island was closed during the concert, thus forcing people to walk; and crossing the bridge they were facing the midnight sun. However, these references may also be read as a symbolic expression, summarizing the consequences of Mandela's magic – the bridging of 'before and after' and the brightness of future goals.

True to the civil ritual genre, this one emphasized moral and ethical commands, and true to the globalizing civil religion to which it belongs, the issue dealt with belongs to the global order – in this case to the fight against HIV/AIDS. The uniqueness of Tromsø, then, relates partly to its ability to *be* global, act globally and take global responsibility. This type of uniqueness needed a front man from outside, and Mandela – as the icon of the qualities celebrated – was a perfect match. Through Mandela, Tromsø became connected to the world; through his speech in Tromsø, its inhabitants were made ambassadors of a global cause; and as a result of their efforts, they became 'honorary Africans' – members of his family and part of his team.

Concluding Comments – Global Scripts, Local Creativity

Ritual events, Durkheim once noted, allow for 'a form of grandiloquence which would be ridiculous in ordinary circumstances', whose ideas 'lose all sense of proportion and easily fall into every kind of excess' (Pickering 1994, 128). A century later, the media has become one site and vehicle of ritual effervescence. Occasions like the Mandela concert allow for a break in the more mundane world of prosaic reporting and a concomitant turn to the world of religion – to hopes and dreams, the impossible imagined and visions celebrated. What people say on such occasions need not be taken at face value, but nor should it be dismissed as irrelevant or

---

11 *Jørgen Hattemaker* is a standard Norwegian term for 'ordinary people'.

without consequences. Mandela has not disappeared from Tromsø. Two years after the concert, memories and pictures still circulate on local media Internet pages, the concert site has been officially re-named 'Mandela-*sletta*'– after lengthy discussions in political forums (and the newspapers); a 'Mandela day' has been established, and events such as 'Mandela-workshops' are being planned.[12] Mandela has also been granted a place in the revised proposal for the Olympic Games in Tromsø 2018, through a 'Mandela Hall', perhaps even a 'Mandela skating hall' on 'Mandela-*sletta*'.

On a broader basis, the concert may have contributed to the more visionary and ambitious tone of recent political discourse in Tromsø, perhaps also to the legitimacy of such discourse among locals. 'New stories' – in order to be acceptable to locals – depend upon some kind of grounding. In this case, 'old' stories were granted new and more positive meaning, mainly based on broader changes in late modern culture. These revisions were literally demonstrated, acted out and experienced and, in contrast to similar stories presented in the Olympic Games proposal, they received no opposition. To be 'against' the concert would imply opposition to Mandela and the causes that he stands for – peace, justice, humanitarian rights, even hope for the future and the possibility of change.

The role of 'Norwegian-ness' is ambiguous in these stories. On the one hand, the 'against all odds' motive positioned 'Norway' as an obstacle to be overcome. Descriptions of landscapes, similarly, draw upon images different from those of the national repertoire. The stereotypical Norwegian landscape is the fjord; it is definitively not the Arctic wilderness portrayed through the concert and the Olympic Games proposal. On the other hand, many of the characteristics ascribed to Tromsø appear as exaggerated versions of national image-making. Norway is known as the 'High North' in the European periphery and has attempted to become known as a humanitarian superpower. Based mainly on peace-making and the turn towards so-called value diplomacy in Norwegian foreign aid politics, this latter type of image-making emphasizes smallness, neutrality and – to quote a critical comment by historian Hilde Waage – a tendency to exaggerate national potential and take on responsiblities 'a couple of sizes too large' (*Dagbladet* 26 April 2004). Tromsø, one might argue, is pursuing a similar route, being even smaller, even further north, even more natural and far more magical. As a way of connecting with a globalizing civil religion, this strategy combines global relevance with local uniqueness, difference and independence.

---

12 Tromsø municipality recently sent out invitations to a 'Mandela-workshop' on 4 June 2007. The invitation linked the workshop to Tromsø's responsibilities as a 46664 ambassador town, listing several global issues and problems to be dealt with, including the environment, HIV/AIDS, education, poverty and gender equality.

References

Bell, C. (1997), *Ritual: Perspectives and Dimensions* (New York: Oxford University Press).

Bellah, R.N. (1967), 'Civil Religion in America', *Daedalus* 96:1, 1–21.

—— (1991), *Beyond Belief: Essays on Religion in a Post-traditional World* (Berkeley: University of California Press).

Beyer, P. (1998), 'Globalisation and the Religion of Nature', in Pearson et al. (eds).

Boer, R. (1997), 'Iconic Death and the Question of Civil Religion', in Re: Public (ed.).

Brosius, J.P. (1997), 'Endangered Forests, Endangered People: Environmentalist Representations of Indigenous knowledge', *Human Ecology* 25:1, 47–69.

Chaffer, K. and Smith, S. (eds) (2000), *The Olympics at the Millennium. Power, Politics and the Games* (New Brunswick: Rutgers University Press).

Chidester, D. (1995), 'A Big Wind Blew up during the Night. America as Sacred Space in South Africa' in Chidester and Linenthal (eds).

Chidester, D. and Linenthal, E.T. (eds) (1995), *American Sacred Space* (Bloomington: Indiana University Press).

Couldry, N. (2003), *Media Rituals. A Critical Approach* (London: Routledge).

Dayan, D. and Katz, E. (1992), *Media Events: The Live Broadcasting of History* (Cambridge: Harvard University Press).

Douglas, M. (1984), *Purity and Danger: An Analysis of the Concepts of Pollution and Taboo* (London: Ark paperbacks).

Drivenes, E.A., Hauan, M.A. and Wold, H. (eds) (1994), *Nordnorsk kulturhistorie. Det gjenstridige landet* (Oslo: Gyldendal).

Fairclough, N. (2003), *Analysing Discourse: Textual Analysis for Social Research* (London: Routledge).

Geertz, A. (2004), 'Can We Move Beyond Primitivism? On Recovering the Indigenes of Indigenous Religions in the Academic Study of Religion', in Olupona, J. (ed.).

Hetherington, K. (1997), *Expressions of Identity* (London: Sage Publications).

Hoover, S.M. (2006), *Religion in the Media Age* (London: Routledge).

Kraft, S.E. (2006a), 'Place Making and Ritual Effervescence. A Case Study of the Nelson Mandela Concert in Tromsø, 11 June', *Temenos* 42:2, 43–64.

—— (2006b), 'Åndenes land. Om Statsbygg, de underjordiske og spøkelser i Nord-Norge', *Din. Tidsskrift for religion og kultur* 3, 1–14.

—— (2004), 'Et hellig fjell blir til. Om Tromsdalstind, Samer, OL og arktisk magi', *Nytt Norsk Tidsskrift* 3–4, 237–49.

Lechner, F. and Boli, J. (2005), *World Culture. Origins and Consequences* (Malden: Blackwell Publishing).

Macnaghten, P. and Urry, J. (1998), *Contested Natures* (London: Sage Publications).

Mathisen, Stein R. (2003), 'Ganning. Mediefortellinger om Sámisk trolldom i dagens Nord-Norge', *Din. Tidsskrift for religion og kultur* 4/1, 20–29.

Olupona, J. (2004), *Beyond Primitivism: Indigenous Religious Traditions and Modernity* (New York: Routledge).

Paulgaard, G. (2006), 'Identitetskonstruksjoner – hvor langt rekker de?' *Tidsskrift for ungdomsforskning* 1, 67–88.

Pearson, J. et al. (1998), *Nature Religion Today: Paganism in the Modern World* (Edinburgh: Edinburgh University Press).

Pickering, W.S.F. (ed.) (1994), *Emile Durkheim on Religion* (Georgia: Scholars Press).

Rappaport, R. (1999), *Ritual and Religion in the Making of Humanity* (Cambridge: Cambridge University Press).

Re: Public (ed.) (1997), *Planet Diana. Cultural Studies and Global Mourning* (Sidney: The Lily-field group).

Roche, M. (2000), *Mega-events and Modernity: Olympics and Expos in the Growth of Global Culture* (London: Routledge).

Selberg, T. (2006), 'Festivals as Celebrations of Place in Modern Society: Two Examples from Norway', *Folklore* 117, December, 297–312.

Szerszynski, B. (2005), *Nature, Technology and the Sacred* (Oxford: Blackwell Publishing).

Tromsø 2014, *Faktaark*, http://www.tromso2014.no (the official Internet page of the Olympic Games proposal).

Urry, J. (2003), *Global Complexity* (Cambridge: Polity Press).

Newspapers

Dagbladet 26 April 2005.

Nordlys 9 June 2005; 13 June 2005; 14 June 2005; 16 June 2005; 18 June 2005; 21 June 2005.

Tromsø 13 June 2005; 18 June 2005.

Chapter 19

Modernity Re-enchanted: Making a 'Magic' Region

Willy Guneriussen

Together we are going to create a ground-breaking sports festival and offer the world a moment of Arctic magic.

Home page of Tromsø 2018, November 2006

A change in symbolic 'climate' seems to be going on in the High North.[1] In recent years, new ways of describing and defining the culture(s), the inhabitants and the landscapes have been established – some of them quite fanciful. Both local and national actors are involved in this process of re-defining and re-constructing various aspects of the identity of the northern area. Some of these re-descriptions seem to imply that Northern Norway is about to become a new, vital and 'dynamic area' in the nation, in Europe and even in a global context. Traditionally, this region has been considered a backward, poor, weakly-developed and mostly pre-modern periphery in Norway, in need of state subsidies and regional development programmes in order to become 'modern'. Such a negative labelling of the region has been typical, and not only by 'outsiders' (particularly representatives from national political, economic and cultural centres). It has also been an important part of the northerner's self-understanding or self-image. People in the north have habitually considered themselves subordinate in many respects. They felt that the modern centre in the south, with all its advanced technology, culture and economic power represented a higher level of development. This 'underdog' identity has been challenged in various ways by northerners, particularly since the 1960s.

I shall focus on recent changes in discourses, metaphors and symbols through which a new northern identity and self-confidence are being expressed. What is being expressed? What kind of cultural identity/identities is/are being constructed through these changes on various symbolic levels? Who is expressing them? How are the area and the people being re-conceptualized, and how are the northern landscapes being filled with new symbolic meanings?

---

1   It is not easy to delineate the 'High North' (*Nordområdene*). This term has recently become the new buzzword in discourses on development in the north. In a new governmental High North strategy (*Nordområdestrategien*) it mainly refers to projects in Northern Norway and the Barents Region.

The main empirical focus of my analysis will be the project of promoting the town of Tromsø[2] as a candidate for the Olympic Winter Games (hereafter signified using the Norwegian abbreviation OL – Olympiske leker) in 2018. The project was originally launched in 2003, with the aim of becoming the preferred national candidate for the games in 2014. As this was rejected by the Norwegian government, a new project for 2018 is now competing for the support of the state/government. I shall refer to the main project documents that have been made public by the official OL organization and also to their presentation in the press and other media. I shall also use the biggest newspaper in Northern Norway (*Nordlys* in Tromsø – traditionally a supporter of the Labour party), as this has become an active supporter of the project. I shall also, to some degree, include references to a big national political project that was launched by the government on 1 December 2006, and which had been announced one year earlier in an official speech at the University of Tromsø by the Norwegian Minister of Foreign Affairs: The High North strategy (*Nordområdestrategien*) – a strategy mainly focusing on economic development relating to the search for and use of natural resources (in particular oil and gas). I shall also refer occasionally to other sources (newspapers, radio, TV, and so on) as they seem to relate to this construction or re-construction of northern identity. All citations have been translated into English.

Re-enchantment – Preliminary Reflections

The descriptions and analysis of the current case will raise some questions as to the validity of traditional modern ideas of cultural and social secularization. In particular, I shall argue that there is something wrong with Max Weber's thesis of disenchantment in modern culture (Weber 1922, 594ff.). If Weber – and Karl Marx before him – had been right, we should by now be witnessing a definite disenchantment of (most aspects of) social life and nature. Their 'diagnosis' implied that modernity was about to strip every social and natural reality of any 'veil' of mysticism, magic and enchantment. Any 'miracles' or 'supernatural' events should be considered quite 'trivial', secular events – the results of normal physical processes that sometimes combine to produce unexpected and statistically unlikely effects (for example, 'unexplainable healing of illness'). Weber's main point was that whatever happened, science (and modern culture) would presume as a matter of course that this was the result of normal, natural causes and laws of nature, even if science at the time was not able to detect the specific causes and laws in every case (Weber 1922, 594). A 'spiritual' concept of nature as a realm of signs, symbols, meaning, hidden intentions and 'supernatural forces' which may be influenced by magic should by now be perceived as being incompatible with the modern scientific mind/culture. At most, it should be perceived as private religious beliefs, not impinging on the more general cultural ideal, official and scientific concepts of nature. The same goes for social reality: as Marx and Engels

---

2   Tromsø is the biggest town/municipality in Northern Norway, located at 69 degrees north on the coast, with approx. 62,000 inhabitants. It is the administrative centre of the county of Troms. It is often referred to as the 'capital of Northern Norway', although this title has no official status and is often rejected by other towns competing with Tromsø for status and investment.

declared in *The Communist Manifesto*, capitalism strips every social relation of any remnants of mystery and sacred meaning, and people are finally forced to look at their social life and their relations '…with sober senses' (Marx and Engels 2001, 13).

Even though science and advanced technology have steadily conquered new dimensions of nature, based on a 'disengaged', disenchanted and purely causal conception of nature, there are many indications that modern culture and actors are not quite done with enchantment. There are ways of conceiving and experiencing an enchanted – or re-enchanted – reality. And we are not (only) talking about private beliefs and subjective world-views. I shall mention two examples concerning relations between modern culture and traditional folk religiosity in the north: the Bishop and priests in the northernmost diocese of the modern (and quite 'secular') Norwegian Protestant Church are now willing to perform blessings on 'disturbed' houses and places – houses and places that are described in folk jargon as being haunted by ghosts or creatures from the 'underworld'. The Bishop insists that this is definitely not 'exorcism' (*Nordlys* 3 November 2006, 6). This modern church will not be associated with (or suspected of) any 'magic'. Traditional magical practices are condemned as superstition. Nonetheless, the clergy are willing to visit such 'disturbed' places, read, sing psalms, and say blessings over/for the place. Priests and lay people report that such practices have worked in various cases. The Bishop concludes that modern priests 'are more open to spiritual dimensions of reality' and are more willing than before to respect various aspects of traditional folk religiosity – particularly Sámi traditions. Maybe it is not very surprising that the church is tempted to regain territories lost to secular forces in modern culture. More surprising (maybe?) are efforts at religious re-enchantment that have affected the most secularized and rational modern institutions – as the following example will testify. The Directorate of Public Construction and Property (Statsbygg) was responsible for the construction of a new Science Centre at the Sámi University College in Kautokeino, in the northernmost county of Finnmark. Before the groundwork started, representatives from Statsbygg and the Chancellor of the University College agreed to sleep out one night at the construction site – eating a traditional Sámi meal and performing old rituals in order to obtain 'acceptance' from the place and its spiritual forces (including underworld creatures) for the location of the building (*Nordlys* 8 June 2006). If any 'disturbance' were detected during the night they would have to negotiate with these forces and possibly make the necessary changes to the project. As no 'disturbance' was experienced, the construction process was able to start as planned. It is not clear whether the actors involved were entirely serious in this performance (see Kraft 2006 for an extensive analysis of the case). However, it is a striking fact that thoroughly rational, modern, bureaucratic and science-based organizations involved themselves in this kind of ritual performance directed at 'spiritual' forces and creatures.

This chapter will focus on another kind of enchantment: what I shall describe as the '*secular re-enchantment*' of society, nature and landscapes, which may be observed in the Olympic project – that is, in a big, public, science-based, modern project. We shall encounter a new, spectacular 'Arctic magic' – landscapes, mountains, sea and fjords that are wrapped in a (quite modern) 'veil' of enchantment. Although we are not observing any ordinary religious revival, a few religious elements are activated in this process – particularly the idea of 'magic'.

Olympic Discourses

I have identified – or constructed – four discourses involved in the Olympic project, based on thematic orientation towards visions, economy, national integration and identity. I shall highlight the symbolic constructions of identity, nature and landscape that may be interpreted from various descriptions of the Olympic project and its presentation or treatment in different media.

In this approach I shall distinguish between *meaning* and *actors' intentions*. The various meanings of symbols, signs, expressions and ways of talking/writing cannot be reduced to, or completely collapsed into, the conscious intentions of particular actors (Phillips and Jørgensen 2002, 9ff.). I shall point to ways of talking, writing and thinking about a phenomenon that will locate it within a larger context of speech, speech acts, texts and value hierarchies, so that it acquires significance, force, 'weight' and direction. When we start to talk or write about a heavily-discoursed phenomenon – like this Olympic project – expressions and arguments become spontaneously entangled in various networks of symbols, other expressions, arguments and positions in different media, often without the actor's conscious and clear intention or wish to do so. The singular expressions and speech acts acquire meanings within these extended networks of established meanings over which the actors have little control. My approach will therefore imply a certain *methodological holism*, without completely denying the effects and importance of individual actions and actors' points of view (Guneriussen 1999, 264ff.). When people choose words and expressions like 'spectacular', 'wild and beautiful', 'magic', 'unique', 'peace and reconciliation', and so on, in the discourses of the Olympic project, these various expressions become entangled in networks of meanings and significance which then feed back in or 'seep' into the intentions of the individual actors. Individual intentions are (partly) constituted through acting within established discourse fields and patterns of symbolic meaning.

Visions

From its first 'spectacular' presentation and tentative formulations in Spring 2003, the Olympic project was presented as a 'visionary project'. The first leader of the official committee for OL-2014 declared that arrangements for the games would be developed by a 'vision-driven, value-based learning organization' and such intentions were included in the first application to the government. The choice of expression does not seem to be a coincidence. It fits widespread trends and fashions in contemporary discourses on organizations (see Røvik 1998) and testifies to something modern, timely, future-oriented and dynamic.

I consider this discourse on visions to be the most inclusive in the Olympic project. In various ways it refers to aspects of the other discourses and seems to constitute a sort of unifying 'wrapping'. Visions are in demand and seem to reflect the spirit of the times. They express a wish to see or construct a superior meaning in different activities and projects – a meaning transcending the ordinary, practical and profane level of action. Part of this 'longing' is due to the perceived trivialization of ordinary politics, where politics is reduced to a plain struggle for positions and barely visible changes in the state budget. The old ideologies that were once so important in giving politics higher goals, heated discussions and more lofty visions have crumbled. Politicians appear more like ordinary bureaucrats. In particular, much of the enthusiasm and emotion-stirring symbolism of the old Labour movement has gone and been replaced by factual, disenchanted, technical and bureaucratic jargon as Social Democrats are, for the most part, in the business of administrating and adjusting a vast and complicated welfare state – not in 'building a new society'.

Against this background, the Olympic project has had an important vitalizing effect on public discourses concerning the development of culture and society in the north, supported by majorities and leaders of most political parties. It has in many ways been established (by the proponents) as the great visionary project for Tromsø and the northern part of Norway – a local, regional, national and even global project that is intended to include all people and every class. It was not to be conceived of as just another 'trivial' (although large-scale) athletic event, with all the practical, organizational, material and economic conditions and consequences this entailed. A representative from the local Olympic committee stated on national TV that the Olympic Games in Tromsø would represent something very 'visionary and vital' (Ulvang, NRK-TV, Dagsrevyen 12 January 2007). It was to become something more, something bigger – a project for comprehensive economic and cultural development in the north; something 'path-breaking, Arctic and magic'; a project that would finally rid Northern Norway of its former subordinated, half-colonial client-status in the nation (Vice Mayor of Tromsø, *Nordlys* 2 December 2006); a project that would be favourable for the development of the nation at large; a project that would render visible (to the nation and the world at large) the spectacular nature in the region and the unique identity of its people and culture; a project for environmental concern and improvement with wide-reaching real and symbolic consequences. And finally: a project that would contribute to peace and reconciliation in the world.

> ...all people, in Tromsø, in the region and in the nation can rejoice at a project that will give us so much in the field of environmental concerns, in economic development, in culture and transport, and – not least – in contributing to peace and reconciliation in the world. (Revold[3] in *Nordlys* 16 September 2004)

For years, Norway has tried to appear on the global scene as a 'peace nation' and peace projects have become important political and symbolic 'attractors'. Norway is responsible for awarding the Nobel Peace Prize. The University of Tromsø is (self)defined as a 'peace university' and a special peace education has been established at a Centre for Peace Studies. A former Chancellor at the university – who initiated the 'peace university' – is now heading the national Nobel Committee and is at the same time an important actor in promoting the Olympic project. The current Chancellor of the (peace) university is also an important contributor to the project. We are witnessing the way in which a wide-ranging net of actors, institutions and symbolic meanings are woven tightly together so as to identify the Olympic project as a visionary peace project – quite in line with Baron de Coubertin's original idea of the Olympic Games.

---

3   Jens Revold is a representative for the left-wing Socialist Party (SV) in the municipality, and an important actor in promoting the Olympic project.

Visions often refer to a higher level of important values and (long-term) goals. The intention is to extend our sight over and above short-term, close and more trivial things, and connect activities to more elevated forms of meaning. 'We' are not just building an ice hockey arena, we are lifting up the entire region to become a modernized area on a par with the rest of the nation; we are making our contribution to a future, peaceful world. There is something very future-oriented and attractive in such visions, and they can heighten the motivation for the different projects. In our time, various 'mega-events' (Olympic Games, other big sporting events, rock concerts, and so on) have seemed particularly relevant, since they have been able to focus a global 'gaze' on the site (see Urry 2003, 82). To be perceived as and to become a (globally) recognized name or 'brand' seems to be a top priority for a post-modern politics of place. Attracting the attention of the global gaze contributes to a kind of re-enchantment of the place and the event. (Cf. Hubbard 1996, 1443ff. on the re-enchantment of urban areas.)

Through this discourse on spectacular visions we are witnessing a kind of secular re-enchantment of culture and nature. Existence is emotionally and aesthetically excited, intensified, filled with a 'higher' meaning and lifted up to a more enchanted or even 'sacred' sphere. It does not have to be 'religious' in the ordinary sense.

Economy

The Tromsø region will of course become very attractive economically if the town is chosen by the International Olympic Committee (IOC) to arrange the Olympic Winter Games. Most spokespersons from the bigger companies in the area have expressed their support for the project, as they find it to be good for the economy in Tromsø and the whole region, contributing towards raising the area's competence in handling various large-scale projects in the future. The scale of investment will be enormous, considering the size of the town. There is potential for making a big profit during the period of construction.

I shall not discuss the controversies about this, nor the long-term economic consequences, but just highlight the perspective of the post-industrial *experience economy* (see Pine and Gilmore 1999) that is implied in this project. Obviously, there are a lot of material and industrial aspects involved. However, the main visions are focused on something more immaterial: the compressed series of spectacular events during the arrangement (just over two weeks), and establishing the region as an attractor for travel and tourism in the global market – highlighting the 'fantastic coastal landscape' and the prospects for people seeking extreme adventure in 'steep mountains rising above the blue ocean'. We may be able to differentiate between two kinds or aspects of experience economy: one focusing on the *aesthetics* of travel and sight-seeing on the part of strangers (the 'tourist gaze' – see Urry 1990), and another variety highlighting the *ecstatic* dimensions of expressive action, extreme challenges, excitement, adrenaline kick, intensity and the momentary. At any rate, the more general focus in this project concerns non-rational aspects of experience, aesthetics and emotional impact: short-term consumption, in other words, not rational, long-term production – although a lot of long-term investment and rational planning/organization has to be made to provide for the main goal of experience and excitement.

National Integration

In various presentations and discussions concerning the Olympic project, the theme of national integration has been recurrent. The background of this specific focus has to do with the history of Northern Norway and a common opinion in the north that the region has not really been fully included and acknowledged to date. The Olympic project is conceived as an instrument to end this subordinate status. Through this project, Northern Norway will be able to raise itself to the level of a fully modern and respected part of the nation. More than that: in public announcements from the local Olympic committee, in various documents and in the final application, the project has been seen as a great contribution to the nation at large – especially as it can present to the world spectacular landscapes and an Arctic wilderness, thereby making Norway more attractive in a global market. Its proponents argue that the world has never before seen this kind of mega event so far north and in such amazing surroundings. The project will mark the end of traditional regional politics – politics of economic support and subsidies for weakly-developed regions. By linking the project tightly to the new and long-term governmental High North strategy, a new symbolic possibility becomes available: to redefine the northern part of the nation as the 'new dynamic region', thereby breaking the old spell of being an insignificant periphery. As the government presented the new national strategy, the Vice Mayor of Tromsø declared: 'This is the end of the client status in the north. Now it is up to us' (*Nordlys* 2 December 2006). The Prime Minister stated in an article that the strategy entailed a co-ordinated national effort including actors throughout the country:

> …this strategy is also about a broad and long-term mobilization of our own powers and resources for the development of Northern Norway. It is not just a project for the north. It is a project for the whole country and the northern part of Europe – and also of importance for the whole continent. (Stoltenberg in *Nordlys* 2 December 2006)

Such arguments acquire symbolic and emotional importance for many people in the north because they seem to signal an important symbolic change, by breaking with the old and ingrained image of a neglected, exploited, inferior and under-developed region. The effort to 'enrol' the Strategy within the Olympic project is clearly expressed in an article by the head of the board of the local Olympic committee, published on the same day as the official launch of the project. The very suggestive title of the article is 'Olympic Games – part of a proactive High North strategy' (Jansen in *Nordlys*, 1 December 2006). The Olympic project is emphasized as a 'natural part' of the national strategy. Being seen as part of a 'proactive strategy' for the north signifies something indisputable good.

A further, very significant contribution to a politics of national integration consists in the inclusion of the Sámi people, who have always struggled hard to be fully acknowledged in the national community. From the beginning, the project has sought to include references to Sámi culture and traditions. The project has succeeded in gaining support from various Sámi officials, the Sámi Parliament and other Sámi organizations, since they consider the project to be an opportunity to attract attention to and strengthen Sámi culture. For the Olympic project, the Sámi element is obviously an attractor – providing an additional, exotic aspect over and above the already fairly exotic northern (or even 'Arctic') region. For Sámi organizations, the Olympic project and the High North strategy are likewise attractors which they try to 'enrol' in their own projects of furthering Sámi culture, identity and interests.

Identity and Landscape

As already indicated, the Olympic project is also about the (cultural) identity of the people(s) in the north. In the previous section I showed that an old, ingrained underdog status has to some extent played a role in promoting the project. The project will show that we are now fully modern, knowledgeable and technologically competent. However, the presentation of this project has underlined that 'we' are not just 'ordinary' modern people, like the people of other modern nations. We are different or *unique* in some important respects – as are the northern landscapes – and this is what will make this Olympic arrangement unique and spectacular. The people are described as open, hospitable and warm-hearted. We are modern, rational and technologically advanced. But we are also characterized by a streak of (creative) 'madness' and 'wildness'. This has been a main point for the Chief Editor of *Nordlys*. Being 'wild' would normally indicate something destructive and uncivilized – not at all compatible with the implementation of this kind of advanced project. But for the Chief Editor, this streak of madness and wildness was (re)described from the beginning as an asset, a kind of primordial power and fantasy that would enable people in this region to create something really spectacular – not just an ordinary sporting event. In various presentations, northerners have been described as sober, but at the same time as visionary. They/we are worldly, down-to-earth and pragmatic, of course, but at the same time in contact with 'magic'. A representative (G. Heiberg) of the International Olympic Committee (IOC) stated on national TV after a visit in Tromsø in 2004 that 'the people of Northern Norway are fundamentally a people of nature'.[4] According to him, this 'fact' should serve as a guarantee that the Olympic project in Tromsø would develop into something out of the ordinary.

These assumptions (if they are to be taken at face value) seem to refer back to exactly the old identity of northerners as not quite modern, not fully civilized. While trying to get rid of the old, negative identity, the proponents of the Olympic project have at the same time used the traditional stereotypes to construct something out of the ordinary – a modern identity that has somehow retained certain aspects of the old, pre-modern identity of a 'nature people' in contact with wild forces and magic, a people living on the edge of civilization, in a border zone that separates the safe and organized modern society and an overwhelming and defiant wilderness. Concepts like this are to be found not only in the regional and national lay discourses on the north but can also be seen in international scientific studies of European identities:

---

4   He used the Norwegian expression 'naturfolk'. In the past this has been translated as 'primitive people', which does not quite fit this context or intentions; in this context it signifies something like 'people who have a particularly close contact with the 'wild' landscapes of the north, and who are fundamentally marked by this relation to nature – by contrast with other modern people'. See also Chapter 18 by Kraft.

'The North represents the natural past, a kind of primordial reference of a people struggling with nature' (Eder 2006, 265).

As indicated, descriptions of landscapes and nature in the north have been important to the promotion of the project. These landscapes are said to be particularly rough, beautiful, majestic and wild – like the people in the north. It is described as an Arctic wilderness. 'Arctic' is a key term that signals a host of positive connotations – clean, white, vast, wild and magnificent. The midnight sun, the 'dark season' when the sun does not rise above the horizon for two months, the winter light, the northern lights (aurora borealis), steep mountains rising above the fjords and the 'blue ocean' are central references in the presentation of these landscapes. Furthermore, all this 'wild and exceptional' nature is highly visible from the centre of Tromsø. In particular, the contrast between majestic nature and modern urban life is highlighted as a condition for constructing a spectacular event (Tromsø 2018, home page, December 2006).

This particular blend of 'wild nature' and civilization, of inhabitants who are modern and rational, though still having traits of a more 'primitive' and 'savage' identity, inhabitants who live in an advanced, high modernity yet still have their roots in the past, makes for a unique identity – an identity that is supposed to be deeply marked by a proximity to the 'Arctic wilderness'.

This kind of interpretation of society, people and nature in the northern region amounts to a (re)enchantment of northern identity and landscapes. The people and the landscapes are elevated above the more trivial and mundane realities of modern existence. They appear as something 'spectacular' and 'magic' – as objects of particular attraction and awe.

Concluding Reflections

In this final section I shall highlight and comment on a few general themes involved in the Olympic project.

Even though there are no obvious ordinary religious motifs involved in the Olympic project, the term 'magic' is often used when referring to the spectacular nature and the planned event. It is not used with any obvious reference to the supernatural, underworld creatures or a spiritual dimension. Why, then, has it become so important? Maybe because it signifies something out of the ordinary, something exceptional, some force that 'electrifies' existence and has an emotional and perceptual impact on us. A downhill race on the 'steep mountainsides rising straight up out of the vast blue ocean' is assumed to incorporate such magical qualities that these will cast a spell, not just on the actual participants but also on the spectators and – maybe even more importantly – on the TV viewers, who will be able to take a virtual part in the events through advanced video production (for example, filming from helicopters as the participants dive down the mountainside). This is not a case of religious magic but a kind of secular, perceptual or experiential 'magic' contributing to a certain 'secular enchantment' of existence that is maybe not so different from certain varieties of religious ecstasy. Magic stands out from an ordinary, highly regulated and relatively safe modern life. It offers the opportunity to establish contact with something outside daily trivialities. It is also, typically, a very friendly and edifying sort of magic – not 'black', dangerous, threatening or destructive. It is a magic suited to modern people seeking extraordinary, strong, positive (and not too dangerous) emotional excitement – a magic cleansed of any 'dark forces' (Alver et al. 1999, 103ff., 132).

It seems to be no mere coincidence that the Olympic project is offering 'Arctic magic' to modern actors who long for something extraordinary. Magic has great market potential in the post-modern experience economy. The northern region ('bordering the Arctic') is seen (or constructed) as being particularly favourable for experiencing modern magic because it is a society situated on the northern fringe of European civilization, and because of its spectacular landscapes.

Wilderness, 'Arctic' wilderness in particular, has been a key term. We have seen how wilderness and magic are symbolically connected. Wilderness has come to be conceived as something very attractive when viewed from within a modern and highly urbanized culture. A change occurred in the conception of wilderness during the nineteenth century, when it was no longer conceived as an ugly, dangerous, unfriendly field of forces not suitable for civilized humans (Cronon 1996, 78). Wilderness stood out as an antidote to 'unnatural' civilization, as a place for freedom and authenticity (ibid., 80). In order to become a healthy, natural human being, uncorrupted by the artificiality of modern civilization, it became necessary to be in touch with wilderness – in nature and in the depths of the soul.

> It is during the nineteenth century that we see a virtual reversal in the symbolism of the natural. The wild, which had once been the epitome of the unnatural, now becomes a natural ideal. (Olwig 1996, 399)

This reversal of the meaning of wilderness is (partly) a consequence of Romanticism and its critique of rationalized and industrialized modernity (Johnson 2007, 23).[5]

The wilderness has become a prime attractor for various forms of tourism – a spectacle for modern spectators, something good and authentic with which to make contact, view or visit. It offers something exciting and out of the ordinary for modern, urbanized citizens. It is wild and it is sublime – but it has somehow become a domesticated, civilized wilderness that can be approached without much danger, with the help of modern technology and organization, or even experienced at a distance through the TV screen. This is a kind and edifying wilderness. As presented by the Olympic project, the northern region appears to have a lot of good, kind and authentic wilderness on offer for modern tourists and spectators.

In modern cultural terms, landscape has generally come to signify scenery, something purely picturesque, beautiful or sublime. It is mainly something to look at – an *aesthetic* concept (Tuan, preface to Olwig 2002, xvii). As Olwig (2002) argues, the concept of landscape was originally conceived of as part of the social, even the spaces of the cities, and part of the body politic. It was primarily habitat, not an aesthetic object. Olwig introduces the concept of '*landscaping*' or land-shaping to highlight the fact that landscapes are always 'scaped' or shaped by practical, social and symbolic activities. So the modern landscape – a landscape to be aesthetically

---

5   See also Tuan (1977 and 1980) and Relph (1976) for more reflections on changing concepts of the wilderness.

apprehended, gazed and marvelled at – is a product of (or shaped by) the increasing distance between urban civilization and nature. It is an object somehow created within the frame of a 'tourist gaze' (Urry 1990). This also seems to be the underlying perspective in the Olympic project – the magical, beautiful and magnificent landscapes are primarily aesthetic objects to be gazed at and admired. They are presented as the most obvious facts of the matter. But as Johnson (2007, 24) has argued, it takes a lot of 'training', socialization and mental habituation to symbolic patterns of meaning to be able 'spontaneously' to perceive various parts of nature as beautiful, sublime, and so on. Olwig has coined a very appropriate term to capture this element of mental habituation: he talks about '*mindscape*' and '*mindscaping*' or -shaping. Just as the landscape is shaped through human activity and interpretation, the mind is shaped in a parallel process. The 'mindscape' of modern people is structured so that they are open, or very receptive, to the aesthetic qualities of landscapes. The ability to absorb the 'tourist gaze' seems to be a widespread modern characteristic.

Landscapes are not only experienced as aesthetic objects. We may also define a kind of *practical* landscape – landscapes as habitat, as spheres of ordinary living. These are not particularly 'enchanted' landscapes. These, then, are not the 'spectacular' landscapes presented in the Olympic project and advertised for the tourist market (see Chapter 17 by Benediktsson).

Landscapes may also be viewed as *expressive*: the forms, structures and processes in a landscape may be read as meaning-embodying signs, as messages directed to human actors – actors who must have specific competence and training to be able to 'read' the various signs in the landscape. This is the landscape of magical influence and communication. This kind of enchanted landscape may be seen in the examples of 'disturbed places' that were mentioned in the first section of this chapter. Such concepts of landscape presuppose certain religious ideas of 'spiritual powers' in nature. The Olympic project has, in one instance at least, come close a 'spiritual' conception of parts of the landscape as well: the Sámi Parliament declared a mountain to be sacred and the plans for downhill slopes on it were abandoned (Kraft 2004).

Finally, I shall hint at another concept of landscape that has been indicated in various presentations connected to the Olympic project – what I shall term *ecstatic* landscapes. These are neither the expressive landscapes housing spiritual powers nor exactly the aesthetic landscapes as objects of contemplation/observation. They are the landscapes of strong emotional sensation, intense experience, impact and adrenaline kicks. They are the landscapes of the extreme sports performers, or adventurers challenging the limits of human capability in demanding surroundings. In some measure these are also available as a virtual reality for spectators and TV viewers.

We have witnessed some quite 'spectacular' efforts to present the landscape and develop new identities for northerners. The new concepts of landscape may result in a new 'Arctic' brand for the region, which may have an impact on the global tourist market. The practical landscapes of daily living in the north will probably not be influenced to a very great extent by this new symbolism. Maybe, however, the landscape approach of the Olympic project reflects tendencies in (parts of) the population – a tendency to view and approach their own surroundings as aesthetic objects, or even as objects of extreme experience and ecstasy.

References

Alver, B.G. et al. (1999), *Myte, magi og mirakel* (Oslo: Pax).

Cronon, W. (ed.) (1996), *Uncommon Ground. Rethinking the Human Place in Nature* (New York: W.W. Norton & Company).

Eder, K. (2006), 'Europe's Borders: The Narrative Construction of the Boundaries of Europe', *European Journal of Social Theory* 9:2, 255–71.

Guneriussen, W. (1999), *Aktør, handling og struktur. Grunnlagsproblemer i samfunnsvitenskapene* (Oslo: Tano Aschehoug).

Hubbard, P. (1996), 'Urban Design and City Regeneration: Social Representations of Entrepreneurial Landscapes', *Urban Studies* 33:8, 1441–61.

Johnson, M. (2007), *Ideas of Landscape* (Oxford: Blackwell).

Kraft, S.E. (2004), 'Et hellig fjell blir til', *Nytt Norsk Tidsskrift* nr. 3–4, 237–49.

—— (2006), 'Åndenes land. Om Statsbygg, de underjordiske og spøkelser i Nord-Norge', *Din. Tidsskrift for religion og Kultur*, 3, 1–14.

Lash, S. and Urry, J. (1994), *Economies of Signs and Space* (London: Sage).

Marx, K. and Engels, F. (1998), *The Communist Manifesto* (London: ElecBook).

Olwig, K.R. (1996), 'Reinventing Common Nature: Yosemite and Mount Rushmore: A Meandering Tale of a Double Nature', in Cronon (ed.).

—— (2002), *Landscape, Nature and the Body Politic* (Madison: University of Wisconsin Press).

Phillips, L. and Jørgensen, M.W. (2002), *Discourse Analysis as Theory and Method* (London: Sage).

Pine, B.J. and Gilmore, J.H. (1999), *The Experience Economy* (Boston MA: Harvard Business School Press).

Relph, E. (1976), *Place and Placelessness* (London, Pion).

Røvik, K.A. (1998), *Moderne Organisasjoner* (Oslo: Fagbokforlaget).

Tromsø 2018 (2006), Official home page of the Olympic Project http://www.tromso2018.no.

Tuan, Y.-F. (1977), *Space and Place. The Perspective of Experience* (London: Edward Arnold).

—— (1980), 'Rootedness versus Sense of Place', *Landscape*, 24:1.

Urry, J. (1990), *The Tourist Gaze* (London: Sage).

—— (2003), *Global Complexity* (Cambridge: Polity Press).

Weber, M. (1922), *Gesammelte Aufsätze zur Wissenschaftslehre* (Tübingen: Mohr).

Index